CIRIA C628

GW01553259

Coastal and estuarine managed realignment – design issues

D J Leggett   Babtie Group

N Cooper   ABP Marine Environmental Research Ltd

R Harvey   Halcrow Group Ltd

CIRIA *sharing knowledge* ■ *building best practice*

Classic House, 174–180 Old Street, London EC1V 9BP
TEL +44 (0)20 7549 3300 FAX +44 (0)20 7253 0523
EMAIL enquiries@ciria.org
WEBSITE www.ciria.org

Coastal and estuarine managed realignment – design issues

CIRIA

Leggett, D J; Cooper, N; Harvey, R

CIRIA C628   © CIRIA 2004   ISBN 0-86017-628-2   RP681

This book constitutes Defra/Environment Agency report FD2413/TR.

British Library Cataloguing in Publication Data
A catalogue record is available for this book from the British Library.

Keywords		
Climate change, coastal and marine, construction management, design and buildability, flooding, ground investigation and characterisation, sustainable construction, sustainable resource use		
Reader interest	**Classification**	
Coastal and estuarine managers and engineers, consultants, environmental regulators, geomorphologists, modellers, consenting authorities, environmental advisers	AVAILABILITY	Unrestricted
	CONTENT	Advice/guidance
	STATUS	Committee-guided
	USER	Coastal and estuarine managers, consultants, environmental regulators and advisers, consenting authorities

Published by CIRIA, Classic House, 174–180 Old Street, London EC1V 9BP, UK.

Summary/Crynodeb

This guidance is intended to disseminate knowledge on design issues for managed realignment projects by providing information on the design and management processes. The authors' aim is to raise awareness and improve understanding of this method of coastal and estuarine management, by sharing experience of managed realignment projects in the UK and by drawing on international knowledge.

There is a diverse range of circumstances in which realignment may take place, so, when applying this guidance, practitioners should consider proportionality, particularly the degree of risk associated with realignment and the level of detail needed in investigations. These considerations should form part of the planning of any managed realignment project.

Managed realignment is a developing area of flood management that draws upon many traditional techniques. As managed realignment techniques improve, there will be opportunities to innovate and to feed such innovation back to the more commonplace approaches to flood management. The guidance will need to be updated as the techniques develop, but for now it provides coastal managers and their consultants with comprehensive information on the design process and help with the practical application of this management option.

Bwriad y canllaw hwn yw helpu rhannu gwybodaeth am faterion dylunio sy'n gysylltiedig â phrosiectau Adlinio Rheoledig drwy ddarparu gwybodaeth a fydd o gymorth yn y prosesau dylunio a rheoli. Nod y canllaw yw codi ymwybyddiaeth ynghylch y dechneg hon o reoli arfordiroedd ac aberoedd a gwella dealltwriaeth ohoni, drwy rannu profiad o brosiectau Adlinio Rheoledig yn y DU a thynnu ar wybodaeth ryngwladol berthnasol.

Mae yna amrywiaeth eang iawn o sefyllfaoedd lle gallai adlinio ddigwydd, felly rhaid ceisio sicrhau cymesuredd wrth weithredu'r canllaw hwn, mae hyn yn cynnwys ystyried difrifoldeb y risg a allai fod yn gysylltiedig ag adlinio a pha mor fanwl y dylai ymchwiliadau fod. Dylid ystyried hyn wrth gynllunio unrhyw brosiect Adlinio Rheoledig.

Mae Adlinio Rheoledig yn faes rheoli llifogydd sy'n datblygu ac sy'n tynnu ar nifer o dechnegau traddodiadol. Wrth i'r technegau ar gyfer Adlinio Rheoledig ddatblygu, bydd cyfleoedd i fod yn arloesol a bydd cyfle hefyd i fwydo datblygiadau arloesol o'r fath yn ôl i'r dulliau gweithredu sy'n fwy cyffredin o safbwynt rheoli llifogydd. Yn y dyfodol felly, bydd rhaid diweddaru'r canllaw hwn wrth i dechnegau ddatblygu, ond ar hyn o bryd ei nod yw bod yn ganllaw cynhwysfawr i'r broses ddylunio, ar gyfer y rhai sy'n gyfrifol am reoli'r arfordir a'u hymgynghorwyr ac fe fydd yn gymorth i'r sawl sy'n ceisio rhoi'r opsiwn rheoli hwn ar waith.

Acknowledgements

Research contractor

This publication is the result of CIRIA Research Project 681 "Coastal and estuarine managed realignment: design issues". It was prepared by Babtie Group, in conjunction with ABP Marine Environmental Research Ltd and Halcrow Group Ltd.

Authors

Daniel J Leggett BSc, MCIWEM – Babtie Group
Daniel Leggett's experience of coastal and estuarine management spans almost 20 years. This has included personal involvement in managed realignment from its strategic base through to practical implementation and scheme monitoring. He has contributed to a wide range of R&D and best practice guidance for the water environment.

Nicholas Cooper BEng, PhD, CEng, MICE – ABP Marine Environmental Research Ltd
With more than 10 years' experience of assessing physical processes and geomorphological issues relating to coastal defences and marine developments, Nicholas Cooper has had practical involvement in managed realignment schemes throughout England. His work has covered initial design considerations through assessment of scheme impacts to monitoring of scheme performance.

Robert Harvey BA/MA (Natural Sciences) – Halcrow Group Ltd
Robert Harvey was project manager of a recent Defra research project on the benefits and applicability of managed realignment throughout England and Wales. He has more than 10 years' experience of environmental and consent issues relating to coastal defences.

Other contributors

CIRIA would also like to acknowledge the contribution of the project team to parts of the report, in particular:

Laurence Banyard	Halcrow Group Ltd
Helen Dangerfield	Babtie Group
Jason Drummond	ABP Marine Environmental Research Ltd
Rosie Elmes	Babtie Group
Natalie Frost	ABP Marine Environmental Research Ltd
Yusuf Kaya	Babtie Group
Nigel Pontee	ABP Marine Environmental Research Ltd
Ian Townend	ABP Marine Environmental Research Ltd
Ray Traynor	Babtie Group
Dave Wheeler	Halcrow Group Ltd
Jackie Young	Babtie Group

Steering group

Following CIRIA's usual practice, the research project was guided by a steering group, which comprised the following.

Chair	Craig Elliott	CIRIA
Attending members	Peter Barham	ABP
	Andrew Bradbury	SCOPAC
	Tim Collins	English Nature
	Alan Inder	Hampshire County Council
	Rod Jones	Countryside Council for Wales

	Paul Murby	Defra
	Terry Oakes	Terry Oakes Associates
	Mike Owen	Environment Agency/Defra R&D representative
	Phil Perkins	Teignbridge District Council
	Matt Simpson	Posford Haskoning Ltd

Corresponding members	Andrew Beattie	Sussex Downs Conservation Board
	Janet Brown	WWF
	Philip Couchman	Chichester Harbour Conservancy
	Andrew Davidson	CADW/Gwynedd Archaeological Trust
	Adrian Dawes	Mouchel
	Pat Doody	National Coastal Consultants
	Rob Jarman	National Trust
	Steve Maslen	Maslen Environmental
	Stephen Midgley	Forth Estuary Forum
	Peter Murphy	English Heritage
	Peter Robertson	RSPB
	Catherine Stradling	Sussex Downs Conservation Board
	Mike Taylor	AAONB
	Phil Tolerton	Defra
	Tim Venes	Norfolk Coastal Partnership
	Tony Weighell	JNCC
	Aidan Winder	Devon County Council
	Phil Winn	Environment Agency
	Stephen Worrall	English Nature/LIFE Living with the Sea

CIRIA managers   CIRIA's research managers for the project were **Elizabeth Holliday** and **Marianne Scott**.

Project funders   The project was funded by:

Defra/Environment Agency joint flood and coastal management research and development programme

English Nature

Countryside Council for Wales

Hampshire County Council

SCOPAC

Teignbridge District Council

Photographs   The following organisations are thanked for their provision of photographic material:

ABP Marine Environmental Research Ltd

Babtie Group

Environment Agency

HR Wallingford

Lincolnshire County Council

Posford Haskoning Ltd

SCOPAC

Contributors

CIRIA and the authors are grateful for the help given to this project by the funders, the members of the steering group, and by the many individuals who were consulted and provided data. In particular, acknowledgement is given to the following:

Alan Brampton	HR Wallingford
Malcolm Bray	University of Portsmouth
Jan Brooke	Environmental Consultant Ltd
Peter Bye	Environment Agency
Kate Carpenter	Independent consultant
David Collins	Defra
Mark Dixon	Environment Agency/Defra
Mark Elliott	Environment Agency
Jon French	University College London
James Garry	Clackmannanshire Heritage Trust
Jim Hall	University of Bristol
Nicola Meakins	Posford Haskoning Ltd
Andrew Miller	Clackmannanshire Heritage Trust
Paul Miller	Environment Agency
Iris Möller	University of Cambridge
Dave Morris	CEFAS
Ruth Parker	CEFAS
John Pethick	Independent consultant
Jane Rawson	Environment Agency
Sue Rees	English Nature
Julie Richards	Environment Agency
Jonathan Simm	HR Wallingford Ltd
Tom Spencer	University of Cambridge
Karen Thomas	Environment Agency
Lajla White	Defra
Richard Whitehouse	HR Wallingford Ltd

Contents

Figures

Tables and boxes

Tables

Boxes

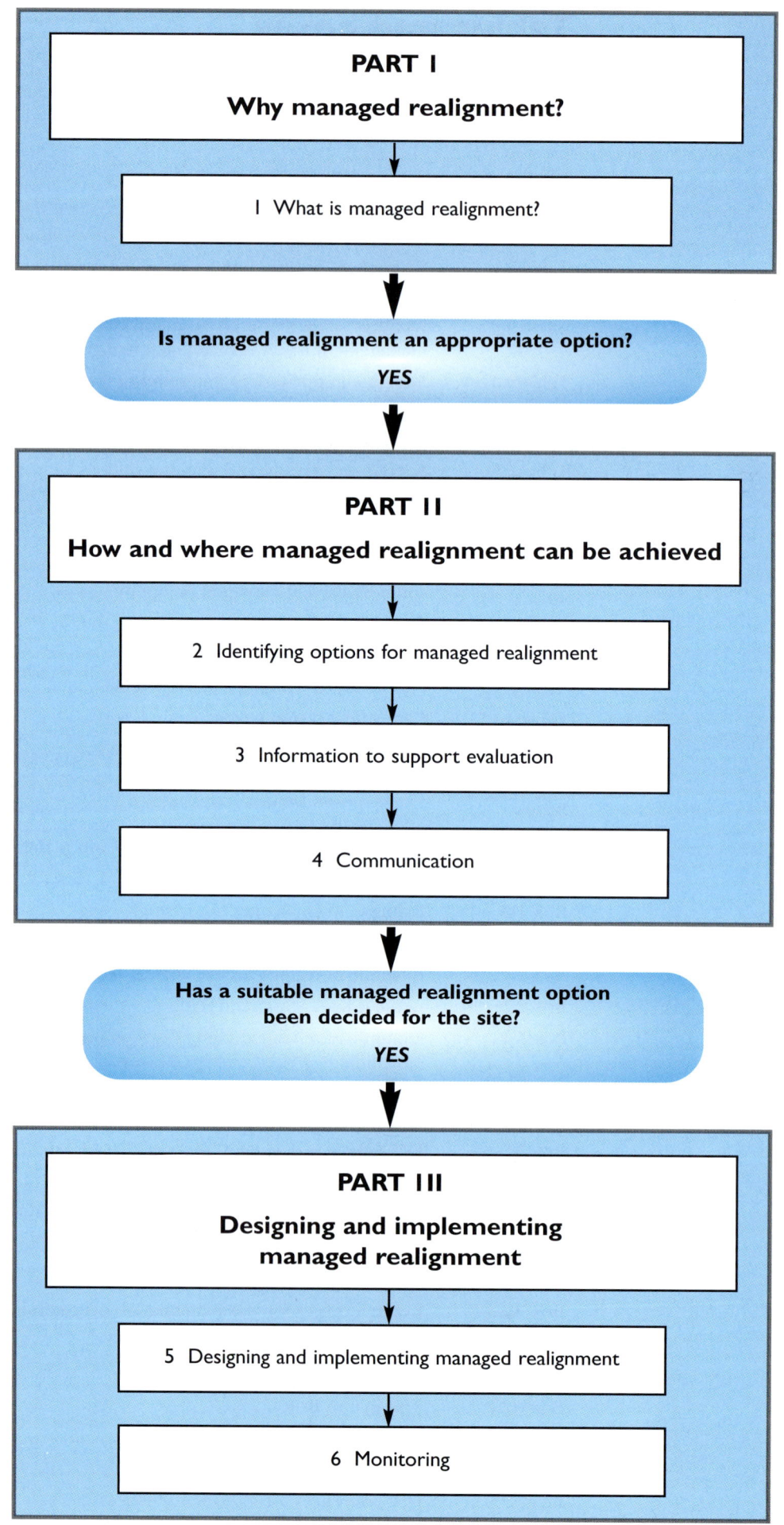

PART I

Why managed realignment?

1 What is managed realignment?

Is managed realignment an appropriate option?
YES

PART II

How and where managed realignment can be achieved

2 Identifying options for managed realignment

3 Information to support evaluation

4 Communication

Has a suitable managed realignment option been decided for the site?
YES

PART III

Designing and implementing managed realignment

5 Designing and implementing managed realignment

6 Monitoring

Introduction

USE OF THIS GUIDANCE

Managed realignment is a relatively new technique compared with wall and embankment design, beach recharge, and the use of groynes and rock structures. The approach to design has been developed over the past decade and it is timely to capture the existing knowledge in this area.

Until now, no comprehensive guidance has been available on the design and construction of managed realignment schemes, although several reports on saltmarshes and habitat creation have been written. The objective of this publication is to help the coastal community to:

- improve the design and implementation of coastal and estuarine managed realignment projects
- facilitate the wider and correct use of managed realignment
- increase stakeholder confidence in managed realignment as a construction option for flood management
- encourage more sustainable design and construction for flood management
- help conserve and enhance natural coastal environments.

The design guide addresses realignment in both low-lying estuarine and coastal situations. Where coastal erosion is referred to this relates to erosion of intertidal areas and coastal features such as beach ridges or dunes.

The book is divided into three parts so that the structure is easily accessible for different users.

PART I explains the objectives of managed realignment schemes. Intended primarily for use by coastal and estuarine managers, it explains the application of managed realignment as an option. It gives a general background to managed realignment and considers success and failure criteria for different stakeholders and how these might have a bearing on site design issues.

PART II discusses whether managed realignment is appropriate for a particular site and explains how it may be delivered. This section is intended for a wider readership, including geomorphologists, managers, consenting authorities, and environmental advisers. Part II explains the decision process that leads to a preferred approach.

PART III provides technical guidance on designing and implementing managed realignment schemes. This part is primarily aimed at modellers, engineers and geomorphologists. The guidance covers the design, implementation, construction and monitoring phases.

A CD-ROM of this publication is included in a pocket on the inside back cover. Readers may find this helpful when searching for specific information.

BACKGROUND

Coastal features such as saltmarshes, shingle ridges and sand dunes act as natural flood defences. They absorb wave and tidal energy at the coast and evolve over time. As coastlines change they may require management or re-creation if they are to continue to fulfil a flood management function. When managing the coast, using natural coastal forms and processes helps ensure sustainability.

As well as providing natural flood defence, the coastline is a vital asset for the UK's biodiversity, including some 1200 nationally protected sites and 180 internationally protected sites, designated for their biological or earth science interest features (JNCC, 1996). The coastline also provides intrinsic landscape value, with more than 1500 km of heritage coast in England [www[1]] and Wales [www[2]] as well as areas of outstanding natural beauty (AONB) and national land assets, such as buildings and farmland. Notwithstanding its value, the coastline experiences pressure from many sources, which may affect the environment and require mitigation or compensation for those impacts. There is, therefore, a need to manage our coastlines in a way that reduces risks to coastal settlements and industries while conserving habitats, biodiversity, geodiversity, heritage and landscape.

Managed realignment can offer long-term sustainable management of coasts and estuaries for a variety of stakeholders and is one of four accepted policies for a flood management strategy. Managed realignment can reduce the pressures of coastal "squeeze" (ie the loss of habitat seawards of flood defences) and offer potential for new habitat creation and re-creation opportunities. Compared with hard defences, the technique may also demand fewer resources and can provide the opportunity for reuse of materials extracted from a site during construction. This can reduce the impact, and on occasions the cost, of construction compared with traditional approaches – for example, by reducing or removing the need for material transport to and from the site. In some circumstances, however, costs might be higher initially, with savings only being realised in the long term through the achievement of a more self-sustaining form of flood management (Owen, 1984; Leggett and Dixon, 1994). For this reason, it is important to consider the whole-life costs as part of a scheme design. It should also be recognised that managed realignment might be applied to achieve land management targets (such as for biodiversity, landscape, public access, archaeological or other social purposes) beyond purely flood management needs, thereby making the cost comparison with other management solutions more acceptable.

Despite the benefits that can arise from managed realignment, over the past decade schemes that appeared to be viable have not all been progressed (Halcrow Group, 2002b). Local communities have rejected some projects that they considered did not meet their needs or where they wished to keep the sea as far as possible from their homes and assets. There has been concern over the impact on landscape and tourism and also a lack of financial priority to progress these projects compared with defending urban areas. The approach may also be rejected because initial designs do not identify sufficient cost savings and the scheme design may over-compensate for the perceived risks and the level of understanding of technical issues involved. Many of these problems have arisen from a lack of available guidance material and limited experience among coastal engineers and managers in designing and implementing managed realignment in the UK, in comparison with more traditional coastal defence options. A result of these weaknesses may be that the potential effects of scheme design options are not fully considered and schemes may need remedial works, or may fail, during their life.

GENERIC LITERATURE REVIEW

Information contained within previous and ongoing literature reviews usefully describes what activities were undertaken on certain schemes, but little has been published on why certain choices were made, what other options were available, or how certain assessments and decisions were undertaken. These details are critical to understanding current practices in engineering design and assessment of managed realignment schemes. Therefore, this report includes information and experience from unpublished sources (the so called "grey" literature) as well as relevant information from published literature.

The literature used for this guidance includes technical reports in support of scheme design or assessments as well as the results of consultation with key individuals associated with several of the existing UK schemes. The review recognises that much has been learnt from the earliest realignments and hence the more recent schemes represent, on the whole, better practice. The review considers schemes that have been implemented previously and also those that have been designed but not implemented, particularly where there are lessons to be learned from the scheme design. The study has tried to ascertain why these schemes were not implemented and to identify if different design (or assessment) approaches might have resulted in scheme implementation.

Glossary and abbreviations

Annex I habitat type(s) — Natural habitat(s) listed in Annex I of the Habitats Directive for which special areas of conservation (SACs) can be selected.

Annex II species — Species listed in Annex II of the Habitats Directive for which special areas of conservation (SACs) can be selected.

Annex 1 birds — Bird species listed on Annex 1 of the Birds Directive for which SPAs are selected.

AONB — Area of outstanding natural beauty.

Appropriate assessment — A self-contained step in the decision-making process required by the Habitats Directive, which must be undertaken for plans or projects likely to have a significant effect on a Natura 2000 site. This includes either effects occurring as a result of a plan or project, or in combination with other plans and projects not directly connected with the management of the site. The purpose is to determine whether or not the proposals would adversely affect the integrity of the Natura 2000 site for the designated species and habitats.

biogeographical region — A region separated from adjacent regions by barriers or a change in environmental conditions that determines the natural geographical range of habitats or species. The Habitats Directive recognises nine regions. SACs are selected in the context of bio-geographical regions.

Birds Directive — Abbreviated term for Council Directive 79/409/EEC of 2 April 1979 and subsequent revisions on the Conservation of Wild Birds. This aims to protect total numbers and diversity, as well as heritage issues, associated with bird migrations.

cannibalisation — The process of eroding sediment from one part of a landform to feed sediment to another part of the same landform in response to change (such as sea level rise). The term is most commonly applied to shingle ridges in "destroying" parts of the existing ridge to allow modification of the landform or movement inland.

CDM — The Construction (Design and Management) Regulations 1994.

CHaMP — Coastal habitat management plans, tested in England, provide a strategic overview quantifying habitat change (loss and gain) over a 30–100-year period. They identify options to prevent future losses of habitat from Natura 2000 sites, and include advice on the necessary habitat restoration or re-creation to provide compensatory habitat for unavoidable change.

coastal squeeze — The process by which coastal habitats are progressively reduced in area, and lose functionality, when caught between rising sea level and fixed sea defences or high ground.

compensatory habitat — Habitat created to offset loss or damage to a Natura 2000 site, to maintain the coherence of the natural network.

The Conservation (Natural Habitats) Regulations 1994 — Legislation to transpose the Habitats Directive into UK law. Also known as the Habitats Regulations.

cSAC — A candidate SAC, before it becomes a site of community importance. It is treated as if it is a full SAC.

dynamic coastline — A coastline that is modifying through physical processes and resulting in changed form. Such changes may be over a wide range of temporal or spatial scales.

EHWS — Extreme high water spring, this has been used in ecological definitions of vertical extent of species and can be taken as a range of the few highest tides of the year. In practice, it is akin to HAT.

EMS — European marine site.

favourable condition — A term used in the UK to describe the desired state of an interest feature at a site level. Condition is favourable if monitoring shows that it meets a series of targets for measurable attributes of the

	feature. The site assessments will be used in reporting on the Habitats and Birds Directives.
favourable conservation status	A key aim of the Habitats Directive. Conservation status is determined by the sum of environmental influences acting on a natural habitat or species throughout its whole range (air, water, soils etc). It is favourable when these influences result in stable or increasing distribution, abundance, and structure or function for habitat maintenance that will continue in the long term throughout the biogeographical region.
geomorphology	The study of the shape of the Earth's surface (the study of sedimentary processes and the evolution and configuration of landforms – in this case coastal and estuarine environments).
Habitats Directive	Council Directive 92/43/EEC of 21 May 1992 on the Conservation of Natural Habitats and of Wild Fauna and Flora.
HAT	Highest astronomical tide. The highest tidal level that can be expected to occur under average meteorological conditions and any astronomical conditions.
hinterland	This is the area of land landward of the coastline. It is normally outside the active coastal or estuarine processes, except in extreme circumstances. This is where most economic assets will be sited.
interest feature	A natural or semi-natural feature for which a Natura 2000 site has been selected. This includes:
	• any Habitats Directive Annex I habitat
	• any Annex II species
	• any population of an Annex I bird species for which an SPA has been designated under the Birds Directive.
intertidal	Zone of seashore between high and low water marks (also referred to as the littoral zone).
LAT	Lowest astronomical tide. The lowest tidal level that can be expected to occur under average meteorological conditions and any astronomical conditions.
LiDAR	A remote-sensing technique using light detection and ranging to measure relative distance, speed or rotation from the sensor. It can be used to pick up solid surfaces or diffuse objects.
LIFE Nature	One of three areas funded under the EU LIFE Financial Instrument for the Environment, set up to co-finance actions aimed at conservation of natural habitats and wild fauna of European interest. Established to support implementation of nature conservation policy and the Natura 2000 network.
managed realignment (MR)	For the purposes of this guide, managed realignment means the deliberate process of altering a flood defence to allow flooding of a presently defended area. This can include alteration to both (semi-) natural and/or man-made defences and involve identifying a new line of defence (where appropriate this will require constructing new defences landward of the original).
MHW	Mean high water. Average high water over a year.
MHWN	Mean high water neap. Average high water at the lowest (neap) tides of the year.
MHWS	Mean high water spring. Average high water at the highest (spring) tides of the year.
MLW	Mean low water. Average low water over a year.
MLWN	Mean low water neap. Average low water at the lowest (neap) tides of the year.
MLWS	Mean low water spring. Average low water at the highest (spring) tides of the year
mODN	Ordnance Datum (Newlyn) in metres. This is the datum of the land levelling system on the mainland of England, Scotland and Wales, and on some of the closer islands offshore. This datum is equivalent to the average value of mean sea level at Newlyn for the six-year

	period 1915–1921. Due to a subsequent rise in sea level, this datum is now about 0.2 m below mean sea level at Newlyn.
morphodynamics	The changes to a shape (eg of a landform) as it is occurring, usually as a result of physical processes.
morphology	The study of shape; in the context of this publication, landforms.
Natura 2000	The European network of classified SPAs and SACs.
operating authority	Body that undertakes flood and coastal defence activities in England and Wales, usually the Environment Agency or a maritime local authority.
PAG	Project appraisal guidance. Defra guidance on the approach to adopt in developing and evaluating flood management projects.
PAR	Project appraisal report. An Environment Agency summary document that presents the business case for proposed projects to be submitted to Defra for funding. It summarises all the relevant information and facts to enable a reader with no knowledge of the project to be satisfied the correct investment decision, in technical, environmental and economic terms, is being recommended.
PPG	Planning Policy Guidance notes cover issues to be considered by planners when evaluating a planning application or planning legislation. Forthcoming revisions will be termed Planning Policy Statements (PPS) and will be supported by guidance documentation.
proportionality	Applying an extent of investigation, implementation, or monitoring to reflect the size (physically and/or in relation to the importance, risks, or functional consequence) of the project/plan/site in question.
quality of life capital	Quality of life capital is a tool for maximising environmental, economic and social benefits as part of any land-use planning or management decision. Promoted by the four agencies (Countryside Agency, English Heritage, English Nature and the Environment Agency), it reflects the Government's integrated approach to sustainable development.
Ramsar Convention	International convention on conservation of wetland habitats and species to which the UK Government is a signatory.
SAC	Special area of conservation. A UK site proposed as a site of [European] Community importance (SCI) designated by a member state through a statutory, administrative and/or contractual act where necessary measures are applied to maintain favourable conservation status.
SCI	Site of [European] Community importance. A site that contributes significantly to the maintenance or restoration of favourable conservation status in the bio-geographic region in which it occurs. These sites are selected by the EC from the list of SACs put forward by member states.
setback	The realignment of existing defences landwards of its existing location, usually within the floodplain.
shoaling	The change in height of a wave as it approaches the shore, in response to water depth change.
site	Defined in the Habitats Directive as a geographically defined area whose extent is clearly delineated.
SMP	Shoreline management plan – used in England and Wales presenting flood and coastal defence policy for a geographic coastal unit (with identifiable coastal processes). Produced by operating authorities under guidance from Defra, SMPs take into account wider environmental and socio-economic requirements when deciding the policy. The revision of SMPs, under the second round, is referred to as SMP2 guidance.
SPA	Special protection area classified under the Birds Directive Article 4.
SSSI	Site of Special Scientific Interest. National conservation designation in England, Wales and Scotland. In Northern Ireland these are called Areas of Special Scientific Interest (ASSI).

TAN	Technical Advice Notes provide guidance for planning legislations in Wales (similar to PPGs in England).
tidal prism	The total volume of water transferred over a tidal cycle (between low and high tides).
United Kingdom	The UK comprises England, Northern Ireland, Scotland and Wales, but excludes the Channel Islands and the Isle of Man.
UK BAP	The UK's biodiversity action plan is an initiative to maintain and enhance biodiversity. Countryside agencies, local authorities, and other organisations, from across all sectors, are committed to achieving the Plan's conservation goals over the next 20 years and beyond. Contains species and habitat plans with targets for habitat creation and restoration.
Water Framework Directive	A European environmental directive setting out a detailed and integrated framework for the improved protection and management of all of Europe's water resources and aquatic environments for each catchment through to the sea. Under the requirements of this directive, most surface water bodies must achieve at least "good" ecological status by 2015.
wave attenuation	The reduction in energy of waves from an increase in friction (for example, from fluidisation of sediment, or by vegetation) or by transformation across the intertidal width.

Organisations

Countryside Council for Wales (CCW)

The Countryside Council for Wales is the Government's statutory adviser on sustaining natural beauty, wildlife and the opportunity for outdoor enjoyment in Wales and its inshore waters. It is the national wildlife conservation authority.

Department of the Environment in Northern Ireland

Statutory government agency in Northern Ireland for environmental and water-related issues.

Department for Environment, Food and Rural Affairs (Defra)

Government department in England with responsibility for Habitats Directive implementation and flood management (flood and coastal defence). Sponsoring department for English Nature and Environment Agency.

English Heritage

Government agency for heritage and archaeological issues in England. The agency aims to ensure the maintenance and care for the historic environment.

English Nature (EN)

Government statutory adviser for nature conservation for England. The agency aims to ensure environmental objectives are included within a framework of sustainable development.

Environment Agency

Government-funded environment protection agency for England and Wales. The lead operating authority and adviser for flood risk management within a framework of sustainable development.

Joint Nature Conservation Committee (JNCC)

The JNCC is the UK Government's wildlife adviser, undertaking national and international conservation work on behalf of the three country nature conservation agencies: Countryside Council for Wales, English Nature, and Scottish Natural Heritage.

Natural Environmental Research Council (NERC)

One of seven UK research councils that fund and manage scientific research and training in environmental sciences including environmental change.

Scottish Environment Protection Agency (SEPA)

Government agency tasked to protect the land, air and water environment in Scotland.

Scottish Natural Heritage (SNH)

Government statutory adviser for natural heritage in Scotland. The agency addresses issues relating to natural, genetic and scenic diversity.

PART I

Why managed realignment?

Part I explains the objectives of managed realignment schemes. This section is intended primarily for use by coastal and estuarine managers and it explains managed realignment as an option. It gives a general background to managed realignment and considers success and failure criteria for different stakeholders and how these may have a bearing on site design issues.

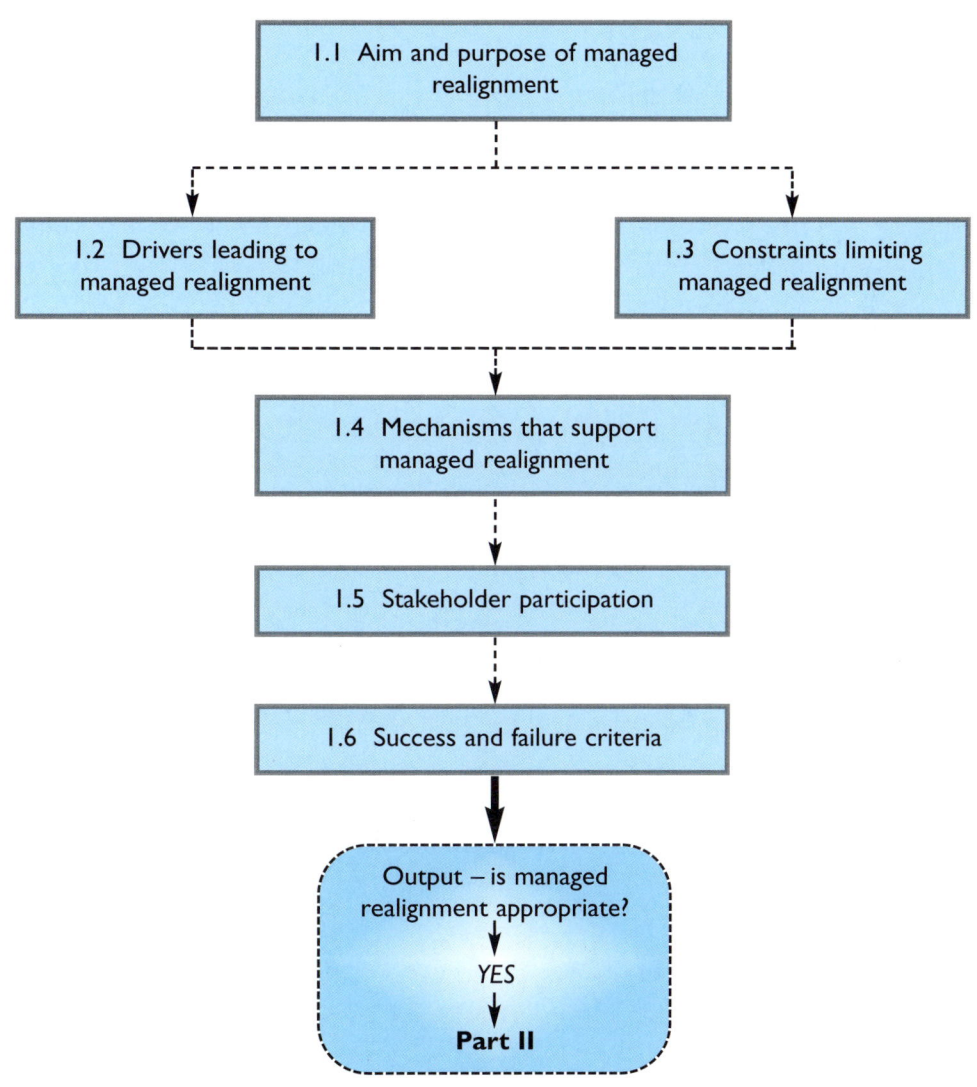

1   What is managed realignment?

Managed realignment means the deliberate process of altering flood defences to allow flooding of a presently defended area. Managing this process helps to avoid uncertain outcomes and negative impacts and to maximise the potential benefits. Managed realignment may take many forms, dependent on the reasons for undertaking it and the techniques, or combination of techniques, applied (Defra, 2003a). Managed realignment can also refer to additional coastal management processes eg controlled retreat of a cliffed coastline, but these are outside the scope of this publication.

The figures below illustrate examples of the different approaches to managed realignment.

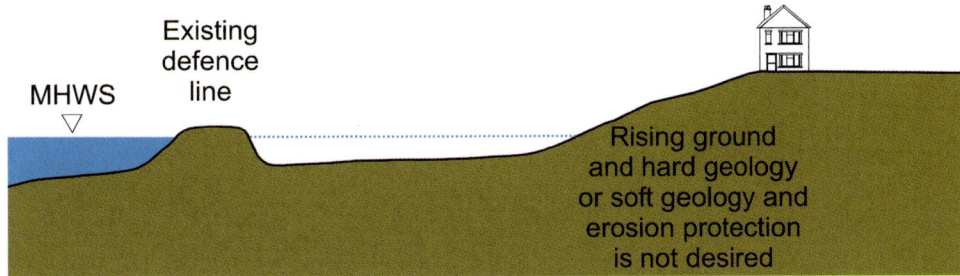

Figure 1.1   *Retreating to higher ground, ie where a line of defence is breached/removed and there is higher topography behind the old line of defence. This allows inundation up to the higher ground and may produce a new intertidal area. A new line of defence may not be necessary*

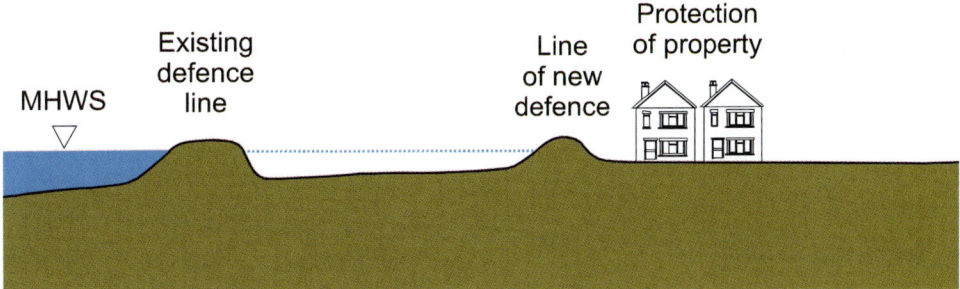

Figure 1.2   *Constructing a set-back line of defence. This can protect property that is landward of an existing defence, or where the ground behind is either lower or not much higher topographically. This enables the defences to be set back and potentially reduced in height*

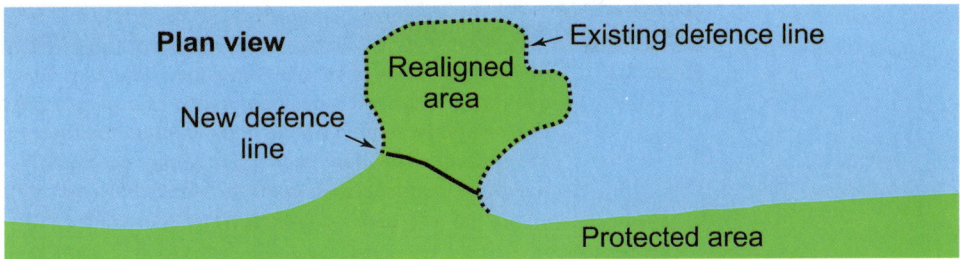

Figure 1.3   *Shortening the overall defence length to be maintained. If the length of the existing defence is large (and thus expensive to maintain in relation to the protected assets), managed realignment can be used to reduce the overall line of defence*

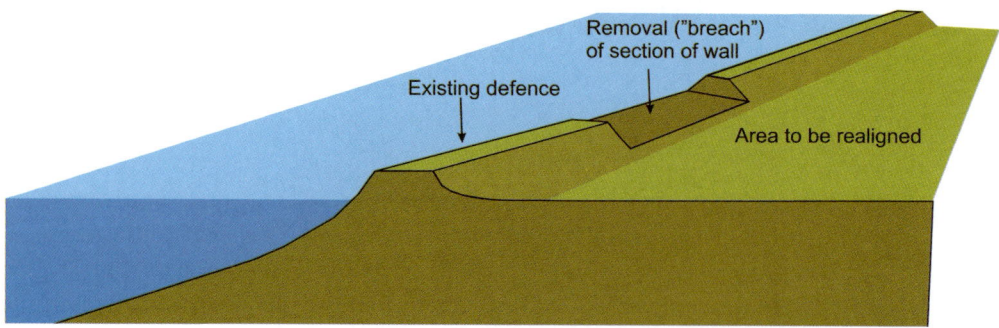

Figure 1.4   *Inundation through sections of defence. A breach managed realignment allows inundation of the land behind through a defined gap. This may be desirable to control inundation of the area or flows from it (and may be armoured if it needs to be maintained for a long time). Simple dug breaches can be cost effective and provide a degree of protection to the landward area while the new intertidal area develops. Pipes with tidal flaps, or other control structures, can be placed through a defence. These can be used to allow an intertidal area to develop and/or gain height (as sediment in suspension settles out) in advance of breaching the defence and are called tidal exchange systems*

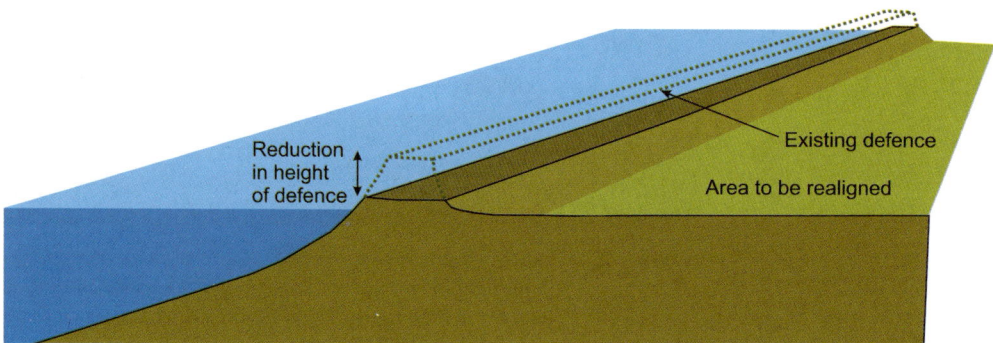

Figure 1.5   *Reducing the height of a defence or entirely removing it. It may be preferable to remove a defence completely (bank realignment) to allow a fully natural system to start to develop immediately. This approach can also avoid scour that may be caused by breaches. The reduction in defence height may be to ground level or to a slightly higher elevation to provide a sill that erodes down gradually and can encourage sedimentation to landward. The height of a defence may also be reduced, rather than removed entirely, to allow land use to continue for at least part of the time with a reduced flood defence level*

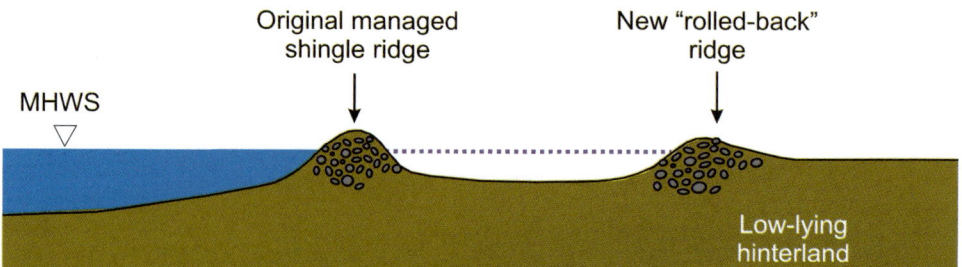

Figure 1.6   *Allowing roll-back of shingle/sand dune defences. This would involve either the controlled or natural movement inland of a shingle ridge or sand dune. The degree of intervention will be dependent on the forces at work and the ability of the system to maintain its function (either ecological or flood defence) while it migrates. In most cases, managed ridges have an artificially high crest level compared with their natural form*

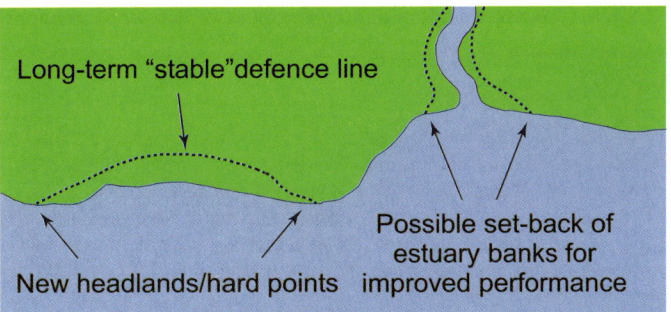

Figure 1.7   *Modifying the morphology to achieve a more sustainable system. The planform of the coast or an estuary might be modified to improve the overall functioning (hydrodynamics) of the system or provide a more sustainable position for the coastline. The nature of change will depend on the situation but might include the creation of a stable bay form on the open coast or the modification of an estuary to move it towards an equilibrium form or to accommodate increased tidal volumes as a result of sea level rise*

Managed realignment should be distinguished from non-intervention or defence abandonment ("do nothing"), in which the process of defence failure is uncontrolled. In some instances, where there is no economic or environmental case for managing the process, abandonment may be the most appropriate solution. The requirements of the Water Framework Directive, however, may mean that managed realignment is used in the future to assure appropriate outcomes for water quality (see Section 1.2.2).

The rest of this section looks at the rationale for managed realignment in terms of:

- the aims and objectives of managed realignment

- the underlying drivers (incentives) for managed realignment

- the underlying constraints (obstacles) to managed realignment.

1.1   AIM AND PURPOSE OF MANAGED REALIGNMENT

It is important to define the aim and purpose for a proposed managed realignment project at the outset. Without this, it may be difficult to determine later if particular success or failure criteria have been met. It is important to investigate all the strategic, opportunistic and legislative issues and to ensure that the intended outcomes are achievable and realistic.

Aim

This is a specific goal of the project ("what are we trying to do..."). Its achievement needs to be defined in measurable terms – for example, a sustainable system, reduced flood risk, or hectares gained through realignment or habitat created.

Purpose

This is the reason why a project is pursued ("why are we trying to do it...") and identifies the benefits and beneficiaries. It should be in line with the aim for the realignment but may present different objectives – for example, reducing costs, improving flood defence performance and sustainability, or creating a new habitat.

Box 1.1 *Examples of the aim and purpose of two realignment schemes*

> **Example 1**
>
> The aim of the scheme is to realign the tidal defence along a 2500 m length by up to 500 m landward of the existing defence. This will create approximately 80 ha of new intertidal habitat. The purpose is to provide a sustainable defence over at least 50 years to a 1 in 200-year standard of protection, which meets technical, economic and environmental criteria. Estuary-wide, the scheme provides an opportunity for significant environmental enhancement and provides compensatory habitat for the loss of intertidal habitat as a result of other schemes within the estuary.
>
> **Example 2**
>
> The aim of the scheme is to realign approximately 5 km of the existing tidal defence to high ground. This will create 400 ha of new intertidal habitat. The scheme forms part of the Environment Agency's strategic approach in the estuary to improve the tidal defence system making the estuary regime more sustainable and providing new habitat.

It is also important at the outset of a project to identify any potential opportunities and/or constraints. In addition to the aims of the scheme, there are likely to be additional benefits that could be maximised through early consultation with stakeholders. The extent of such opportunities might be constrained, for example if it is a Natura 2000 site some activities may be unacceptable for the functioning of the system or the disturbance they cause.

1.2 DRIVERS LEADING TO MANAGED REALIGNMENT

A review of managed realignment by Defra and the Environment Agency (Halcrow Group, 2002b) identified the main drivers for realignment, as seen by a range of practitioners and other stakeholders. The relative importance of these factors varies between schemes and they are presented below in alphabetical order. More than one driver may apply to a proposed realignment.

1.2.1 Coastal and flood management

Managed realignment can be an effective way to provide flood risk management in terms of:

- reduction in flood risk or potential for reduction in whole-life costs of providing flood defence (capital and maintenance costs reduced to present values). The principal benefit would be to the organisation(s) charged with managing and funding flood defence and to the people protected by the enhanced defence

- potential to change the hydrodynamics of an estuary or coast so as to reduce the risk of flooding at another location or to improve the functioning of the hydrodynamic and sedimentary system. The principal benefit would be to the organisation(s) charged with managing and funding flood defence, with less tangible benefits to society as a whole

- ensuring long-term sustainability of defences, for example by increasing natural flood and storm buffering capacity

- reducing the costs, particularly where it is no longer economic to defend land or where realignment enables the defence line to be moved to naturally higher ground

- avoiding abandonment where absence of management may create greater problems

- managing the effects of sea level rise by moving defences landward to less exposed positions and providing capacity for estuaries and coasts to adapt to higher (and lower) sea levels

- taking advantage of the intertidal area to reduce waves and thus reduce the capital and maintenance costs of realigned flood defences
- maintaining the health of the beach (including ridges or dunes) as the first line of coastal defence
- reducing the tidal and wave energy, and therefore erosion at the open coast through creation or restoration of a tidal delta.

1.2.2 Environmental benefit

Managed realignment offers the potential to mitigate the effects of previous reclamations and of climate change (in particular, sea level rise). In combination with retention of existing defences, these effects may cause habitat loss ("coastal squeeze"). The principal beneficiary would be "the nation", but it is also essential for those organisation(s) charged with managing habitats, and the organisation(s) responsible for maintaining past structures (that may be constraining landward migration of habitats).

Figure 1.8   *Creation of habitat and amenity. Managed realignment can create habitat and provide a new amenity through improving public access*

Many governmental and non-governmental organisations wish to mitigate changes in habitat and to restore, conserve and/or enhance coastal and estuarine environments by creating new areas of habitat. This will include general societal benefits reflected in government targets for biodiversity, landscape, heritage, public access and environment.

The most apparent environmental benefit from managed realignment is often intertidal habitat creation. Usually this is in the form of saltmarsh and/or mudflat, but it may also allow space for roll-over (either naturally or mechanically) of shingle ridges or sand dunes, often in association with saltmarsh and intertidal flats. Restoration and enhancement of these habitats provides a key contribution to delivering the targets set by the UK Biodiversity Action Plan and Defra's High Level Targets (Defra, 1999).

Managed realignment can benefit the environment by providing:

- potential for development of a type of habitat of greater biological diversity or sustainability than the existing habitat or land use. This principally benefits the natural environment and organisations charged with managing habitats, but also provides less tangible benefits to society as a whole

- potential for a habitat to compensate for change to other habitat of the same type elsewhere, for example as a result of a land development. The principal benefit would be to the developer, although some benefit may be derived by the organisation(s) charged with managing habitats or flood defence

- potential enhancement of recreational value and access (for example, wildfowling, bird-watching, dinghy-sailing and walking); improved access to these environments can be achieved with a realignment project

- enhancement for fisheries (for example, creating nursery areas for young fish, expanding the food resource available in an estuary and providing areas for shellfish beds)

- a valuable nutrient and pollution sink, improving water quality and reducing the undesirable effects associated with eutrophication. This will become increasingly important to meet the requirements of the Water Framework Directive

- enhancement of landscape by returning it to a more natural state or removing man-made features. This may be particularly relevant in AONBs, national parks or heritage coastlines

- the development of a sustainable estuary shape to contribute to flood management.

Water Framework Directive

Managed realignment offers potential benefits to the UK in meeting its obligations under the EU Water Framework Directive, including the requirement to restore water bodies to good ecological status by 2015. As the needs of the Water Framework Directive develop there may be the opportunity to use realignment areas to maintain the condition of water quality and the functioning of morphology within a coastal or estuarine system. Re-establishing the ecological integrity of intertidal habitats may also be an important driver for managed realignment.

With respect to the design of managed realignment schemes, the baseline for the directive is to establish *natural* ecosystems, which may have implications. For example, a managed realignment design involving a sluice gate to allow water in and out of an area to enable saltmarsh to establish, may satisfy the requirements of the Habitats Directive, yet may not satisfy the Water Framework Directive unless it can be proved that the development of more natural systems would either be not technically viable or disproportionately costly.

1.2.3   Funding

The availability (or non-availability) of funding may affect a decision to opt for managed realignment. There are three principal sources of relevant funding.

Flood and coastal defence funds

Flood and coastal defence funds are available if economic, environmental and technical criteria are met. In England and Wales, these are administered by Defra and the Environment Agency while in Scotland and Northern Ireland they are the responsibility of the devolved administrations. If managed realignment at a given site is better able to meet the necessary criteria than alternative approaches, it may be eligible for funding. Funding to purchase land over which managed realignment is to take place is available only in certain circumstances, as outlined in Defra's policy paper *Managed realignment: land purchase, compensation and payment for alternative beneficial land use* (Defra, 2003a). The paper states:

> *The purchase or long term-lease of land is a legitimate charge on the flood management budget when:*
>
> - *it is required for the building of set back defences, i.e. the land under the 'footprint' of the new defence*
> - *it will contribute to the performance of the new defences, e.g. where the wave attenuation of a stretch of land seaward of a new defence will allow smaller defences to be built*
> - *it is required for habitat creation to offset the impact of new defences or sea level rise on a Natura 2000 site.*

In some cases, operating authorities have offered to buy land adjacent to estuaries from willing sellers, to be used for intertidal habitat creation as part of a strategic approach to flood management.

Agri-environment payments

Agri-environment payment schemes include:

- environmentally sensitive areas throughout the UK
- Countryside Stewardship administered by Defra in England
- Tir Gofal administered by the Countryside Council for Wales
- Rural Stewardship Scheme operated in Scotland
- Countryside Management Scheme in Northern Ireland.

These can provide set payments for creating intertidal habitat on grassland or cultivated land, for example, for a period of 10 years in the case of Countryside Stewardship. Payments are also available to meet capital costs incurred by private landowners in setting up a realignment scheme, but typically these only meet a proportion of the full costs. Defra is reviewing its agri-environment schemes and it is understood that the intertidal options are likely to be retained but payment arrangements are under review.

Private capital/non-governmental funding

Funds may be available from private companies or non-governmental organisations for the creation of habitat areas for a variety of reasons. These may involve "green" public relations, habit to replace those lost as a result of development, or creation of nature reserves. In some cases, such habitat creation is driven by legislative requirements (see Section 1.4.3).

Private developers have bought land at market values to achieve mitigation for change caused by new development. Non-governmental organisations (NGOs) have also purchased land privately for habitat creation to support their conservation remit. Some have also changed management of land-holdings to enable realignment on their land.

PART I

1.2.4   Legislation

Flood and coastal defence in England and Wales is largely permissive and not a legal requirement. In general, operating authorities are not required to build or maintain any particular line or standard of defence, or indeed provide defence at all (except in very limited circumstances where local historic legal agreements exist). Decisions about where to build and maintain defences, and at what standard, are based on economic, technical and environmental criteria. Since there is no legal requirement to provide or maintain defences, abandoning existing defences where they do not fulfil the criteria is a real option.

Defra grants to support the cost of flood management projects undertaken by the Environment Agency are described in the Memorandum Relating to Flood Defence Grants to the Environment Agency (June 2003). The memorandum derives from the Water Resources Act 1991 and the Land Drainage Act 1991 and should be read in conjunction with the Land Drainage (Grant) Regulations 1967. Separate memoranda on grants cover local authority and internal drainage board flood defence projects, and works to counter coastal erosion under the Coast Protection Act 1949.

The following pieces of environmental and planning legislation can act as drivers for managed realignment.

The Habitats Regulations

The European Union Habitats Directive (Council Directive 92/43/EC on the conservation of natural habitats and of wild flora and fauna) was incorporated into UK Law through the Conservation (Natural Habitats, &c.) Regulations 1994. The Habitats Regulations require that a plan or project that might adversely affect the integrity of a classified special area of conservation (SAC) or special protection area (SPA) may only go ahead if there is no alternative, if the scheme is in the over-riding public interest, and if compensatory habitat is provided to replace what will be lost. The UK Government also considers that the interest of sites designated under the Ramsar convention should be treated in the same way.

Projects that entail deliberate reclamation (such as port construction or waterside development), dredging (possibly leading to intertidal erosion), or holding an existing coastal defence line under rising sea level (possibly leading to coastal squeeze) can cause change to intertidal habitats. Where such habitats are qualifying features in an SAC or support qualifying species in an SPA, an appropriate assessment may deem to have an adverse effect on the integrity and functioning of the site. This can lead to adoption of managed realignment in one of two ways:

* as an alternative to holding the existing line of defence; this may help to avoid an adverse effect on site integrity caused by holding the line

* if there are reasons of over-riding public interest, and there is no alternative to the plan or project, compensatory habitat may be required. This could be met by landward realignment at another location.

Where an existing coastal defence prevents saline inundation or erosion of an internationally designated freshwater or terrestrial habitat, the Regulations might then act as a constraint to managed realignment.

Shellfish Waters Directive

In areas designated under the EU Shellfish Waters Directive, there are limits on, among other things, increasing suspended sediment concentrations above background levels for more than a defined period. In some cases, therefore, the directive might constrain a managed realignment initiative.

Town and Country Planning Act 1990

Where a coastal defence would require planning permission (see Section 2.6), other planning considerations, such as those in statutory local plans, may lead to managed realignment if they preclude a hold-the-line option – because of negative effects on landscape, wildlife, geology or amenity, for example. Such considerations may be of particular importance within areas such as national parks, areas of outstanding natural beauty, heritage coasts and other areas designated as being of special landscape and/or amenity value. This could occur, for example, when a primary coastal defence is provided by a mobile beach, shingle ridge or sand dune that is an important landscape, geomorphological and amenity feature. These factors may lead to a decision that the coast should be allowed to behave naturally, for example to maintain the width of beach, and the main line (or area) of managed defence provided to landward of it.

Wildlife and Countryside Act 1981, as amended by the Countryside and Rights of Way Act (CRoW) 2000

The Wildlife and Countryside Act, as amended, creates a duty for public authorities ("Section 28 Authorities") to conserve and enhance sites of special scientific interest (SSSIs) and protect certain species, when exercising their functions. Although less stringent than the requirements of the Habitat Regulations, this duty applies to a wider variety of sites and to a wider variety of classified interests within them, for example incorporating geological and geomorphological as well as biological features. It may act as a driver for managed realignment where the conservation of SSSI interest features would benefit from such an approach.

1.2.5 Navigation

In some cases, managed realignment may enable training structures to be removed from the mouth of tidal rivers by increasing the tidal prism sufficiently to make the estuary mouth self-scouring. This could improve the navigability of the channel and save on future costs of maintaining navigational structures, as well as achieving a more sustainable estuary configuration. The rise in sea level may alter flow patterns and velocities as well as morphology and bathymetry of the system, changing the navigation. It may be possible to counter any negative changes through strategic realignment; such an approach would require careful investigation to provide reassurance that the new regime would be workable. Consideration should be given to whether the realignment will itself cause change in the system; this would be of particular importance for larger vessels where depth might be reduced and for smaller vessels where flow patterns and velocities might be hazardous. Potential benefits would need to be weighed against the uncertainty in outcome and the potential drawbacks to navigation.

1.3 POTENTIAL CONSTRAINTS LIMITING MANAGED REALIGNMENT OPPORTUNITIES

Defra/EA's review (Halcrow Group, 2002b) identified a number of constraints affecting opportunities for managed realignment. The relative importance of each factor varies between schemes. They are presented below in alphabetical order.

1.3.1 Consents and legislation

Consents generally needed for managed realignment include those under:

- Town and Country Planning Act
- Coast Protection Act
- Food and Environmental Protection Act
- The Environment Impact Assessment Regulations
- Land Drainage Act
- Water Resources Act
- Flood Defence (Land Drainage) Bylaws and Sea Defence Bylaws
- Highways Acts.

These are discussed in detail in Section 2.6. The number of consents required and the complexity of the overall consent process can be a disincentive to adopt a realignment option. There may also be local arrangements in place that provide a legal basis for protection. These are not widespread, but it is essential to investigate them. For example, local byelaws or ancient rights may apply as well as legally binding agreements for protection of land and property. Where sites of special scientific interest (SSSIs) are involved, consent may also be needed from the country agencies for nature conservation (English Nature, the Countryside Council for Wales or Scottish Natural Heritage) under the Wildlife and Countryside Act 1981 (as amended).

In addition to providing a driver for managed realignment where intertidal habitat creation is required, the Habitats Regulations can also represent a constraint where the effect would be to turn protected terrestrial or freshwater habitat into intertidal areas. The country agencies for nature conservation can provide guidance on these requirements. Unless deemed "necessary for the management of the site for nature conservation", it must be ascertained whether a proposed realignment would adversely affect the integrity (ie structure and functioning), in habitat, processes and ecology terms, of a (whole) SAC, candidate SAC or SPA, before consent can be given. If it is considered there would be a "likely significant effect" then an appropriate assessment will be needed. Similar requirements may be needed for European marine sites (EMSs) where there is an impact offshore from the realignment scheme. For some EMSs, there may be management schemes, which place a binding obligation on relevant authorities and will thus have certain requirements.

If it may adversely affect site integrity, realignment can go ahead only if there are no alternatives and if there are reasons of over-riding public interest. Where this can be demonstrated, the process of gaining consent under the Regulations can be very time-consuming, as it involves referral to the Secretary of State.

Managed realignment generally requires planning permission, so planning policies may be a constraint. The planning issues are often complex, particularly where realignment is being introduced to create compensatory habitat and if the development and the compensation sites do not come within the jurisdiction of the same planning authority.

Landscape issues may also constrain managed realignment where the character of an estuary or section of coastline will be significantly modified. The landscape might be designated as an AONB or perceived to be important to heritage or local economies in its present form – for example, associated to tourism.

Defence realignment often entails footpath diversion (or possibly extinguishment and creation), as many footpaths run along the existing defence alignment. This requires consent from the local highways authority (county or unitary council); in some instances, obtaining this consent may represent a significant constraint. Public and private access routes have been identified as one of the most constraining factors on managed realignment projects (Environment Agency, pers comm, 2003). Similarly, the presence of services such as pipelines and cables may also be a constraint and require consent or considerable cost for re-routeing.

1.3.2 Environmental issues

Environmental issues that may constrain managed realignment schemes include:

- nature conservation
- landscape and historic environment
- agricultural land use
- recreational use of land or sea
- navigation
- water resources or quality.

Each factor needs to be considered within the context of environmental impact assessment and may have particular requirements as a result of any assessment (see Section 2.5.1). Projects that have been constrained, to date, have been as a result of nature conservation issues (see below). However, concerns over archaeological damage, landscape changes and loss or alteration to footpaths have all been raised in recent projects and/or caused modification in the design process.

The change in existing terrestrial or freshwater habitat (including, for example, reedbeds, lakes and grazing marshes), whether or not protected by the Habitats Regulations, is often seen as a constraint to realignment. There may also be actual or perceived difficulties and uncertainties relating to the feasibility of creating new intertidal habitat or moving ridges or dunes in a realigned area where a very specific habitat is desired. It is necessary that a balanced view be taken of the benefits and disbenefits. Even where the creation of high-quality habitat may not be certain, or even possible, the restoration of a more "functional" environment may still be a reason to proceed and may provide long-term habitat.

The management of gravel or dune ridges may have allowed limited inland migration of these features over time, but the ridge may now be located in an unsustainable position, from which catastrophic breach and seaward drawdown of sediment may occur. In these circumstances, it may not be possible to maintain the ridge by allowing natural processes to be revitalised, and physical reprofiling may be necessary. Natural roll-back may then occur from this more natural, sustainable position.

Figure 1.9 *Shingle ridges may suffer seaward drawdown of sediment if not allowed to roll back inland under natural processes*

It is possible that valuable habitat adjacent to or alongshore of realignment sites may be changed through factors such as increased tidal prism, interception of non-cohesive sediments or increased flow velocities and additional scouring effects. Such change needs to be recognised and accounted for when evaluating the overall net gain or loss. In some instances, an adverse effect on site integrity may be identified, which, if it cannot be designed out, may stop a realignment scheme from proceeding.

Some realignment schemes have encountered opposition prompted by concerns that they may cause redistribution of sediment or stronger currents and thereby damage shellfish beds. These effects can be difficult to predict accurately in advance (set against natural variability), but they can be monitored and planned for when implementing a scheme.

Another constraint may be the presence of contaminated material within or landward of the existing defence. An example would be the presence of an active or former landfill site in an area that might otherwise be considered for realignment. The cost of removing such materials is likely to be expensive and, unless the quantities involved are small, holding the existing defence line is likely to be the only acceptable option in the short to medium term. There is also some concern that water could be polluted through the release of farm fertilisers, herbicides or pesticides from new setback sites. Some contaminants maybe rendered acceptable through effective management of the site before realignment, for example by stopping the use of herbicides and pesticides or allowing the sun (UV) to break down complex compounds in advance of inundation.

1.3.3   Funding and financial compensation

Realignment funding needs to cover the design process, obtaining consents and licences, consultation, construction and, where relevant, management of the site. Land purchase can account for a substantial part of the total cost where it is necessary to achieve realignment or where it is appropriate to compensate landowners.

If certain criteria are met, Defra flood management funding for land purchase is available for managed realignment (see Section 1.2.3). In practice, the Defra criteria can be fulfilled by many proposed managed realignment schemes, particularly where they are undertaken as part of a strategic programme with objectives that include providing replacement habitat under the Habitats Regulations. There would be no funding available from Defra, however, if compensatory habitat is required due to changes caused by private development. Where there is a legal obligation to create compensatory habitat, it is expected that the developer will fund the associated costs, whether this is a flood defence operating authority supported by Defra funding or a private company reliant on its own resources.

Saltmarsh creation may also be eligible for payments under English, Welsh, Scottish or Northern Irish agri-environmental schemes or European Union LIFE funding, but not where the purpose of defence realignment is to deliver statutory compensatory habitat.

Arrangements to buy land for intertidal habitat creation have proved popular in some locations. In others, however, property owners have not wanted to relocate and have considered that such disturbance to their existing way of life should be compensated over and above market value. The current limit of 10 years for stewardship payments can also make it unattractive to landowners. Under present legislation, which is based on flood defence being permissive, there is no right to protection and this may contribute to community opposition to managed realignment. An alternative to managed realignment may be abandonment of the defence, for which neither land purchase nor compensation would be available. During consultation and discussion, it is important that stakeholders are aware of this consequence.

1.3.4 Opposition from the community

Many projects have encountered a lack of community support for managed realignment, at least at the outset. This may be related to:

- loss of land having, or perceived as having, high property value, development potential or personal value, often associated with strong emotional ties

- the perception that loss of "hard-won" land is a retrograde step

- worries about adverse environmental effects from water close to communities (for example, if it became stagnant or malodorous)

- concerns about amenity in terms of loss of assets (such as a beach) or, conversely, increased pressures from additional visitors (by sea or land)

- loss or change to economics of agricultural land

- lack of understanding of the "whole picture", possibly due to insufficient consultation or community participation in a project

- the belief that current funding arrangements do not compensate for their losses under managed realignment

- a perception of being unprotected from flooding (if no new counter walls are needed) or that the level of protection has decreased

- a perception that a hard defence provides absolute protection.

Many of these concerns derive from the perception of the sea as a powerful force that should be kept as far away as possible, rather than something to manage to help reduce the risks to people and property in the long term. The perception may be flawed, as a hard defence only provides defence up to a certain standard and may give a false sense of security. Public perception requires attention in any project. Consultation and participation provide an opportunity to educate a community on the options and the risks associated with them (see Section 1.5).

1.4 MECHANISMS THAT SUPPORT MANAGED REALIGNMENT

While the drivers and constraints provide the underlying rationale for managed realignment, the context in which it is achieved may also be classified according to whether it is strategic, opportunistic or required by legislation.

1.4.1 Strategic managed realignment

Managed realignment may be termed "strategic" when it arises from the implementation of plans and policies, typically covering some tens of kilometres of shoreline and addressing a variety of needs within that plan area. This is a "top-down" approach, where sites are selected according to a combination or hierarchy of criteria, such as the sustainability of the morphology, uneconomic nature of existing defences and site suitability for habitat creation. Practical issues such as the willingness of landowners and other stakeholders to participate are also an important factor in site identification within the overall context set by the strategic vision.

A strategic approach to managed realignment is often seen as a desirable way to maximise benefits and overcome potential constraints. In England and Wales, shoreline management plans (SMPs), coastal defence strategies, biodiversity action plans (BAPs) and coastal habitat management plans (CHaMPs) are the principal tools for strategic planning and provide the means to identify opportunities for managed realignment and their required extent. The first generation of SMPs focused on the open coast, and inclusion of estuaries – where managed realignment might be an option – was varied. The second generation of SMPs should address this shortfall. Coastal defence strategies do not cover the whole coastline, and CHaMPs focus on specific and limited complexes of SAC/SPA/Ramsar sites. Catchment flood management plans (CFMPs) are being developed and introduced across the whole of England and Wales and will cover both estuaries and "tidal rivers" where gaps have been left by SMPs and coastal strategies. The Water Framework Directive requires river basin management plans (RBMPs) that cover the whole water system from rivers to the sea. RBMPs will set out, where necessary, how favourable (ecological, geomorphological or water chemistry) status can be achieved through management of the environment, which might include the use of managed realignment. Some organisations have also carried out specific investigations to identify potential realignment sites – for example the RSPB identified nearly 10 000 ha of land where realignment could meet multiple criteria (RSPB, 2000).

Shoreline management plans

The whole of England and Wales and parts of Scotland are now covered by shoreline management plans (SMPs). These documents focus principally on the open coastline (ie they often, but not always, exclude estuarine shores) and where necessary also consider estuary-coast process interactions. The SMPs contain the strategic policy and key information of relevance to obtaining a baseline understanding of physical processes and morphology in the vicinity of a managed realignment scheme, particularly if the scheme is located within a coastal setting (see also Section 3.2.4). They provide a large-scale assessment of the risks associated with coastal processes and present a policy framework to reduce those risks to people and the developed, historic and natural environment in a sustainable manner (Defra, 2001). SMPs are non-statutory but set coastal defence policy for the coast. The options defined in new draft procedural guidance (Defra, 2003b) are "hold the line", "advance the line", "no active intervention" and "managed realignment".

A recent review of the role of SMPs in relation to managed realignment (Halcrow Group, 2002b) found that managed realignment was proposed in 39 (3 per cent) out of about 1100 management units in England and Wales. Of these 17 (44 per cent) had been or were being implemented. A further 16 sites were identified that have been or are being implemented but were not included in SMPs. These statistics suggest that the first generation of SMPs has been of limited effectiveness in bringing forward managed realignment. Natural processes, particularly the dissipation of wave energy in the intertidal zone, were seen as a significant driver for realignment in about half of the cases examined. In many instances, however, the adequacy of available natural process knowledge appears questionable and political considerations seem to have influenced the outcome. The availability of better information over time, such as the outputs of Defra's Futurecoast initiative for England and Wales and the findings of CHaMPs, will have a bearing on the next revision of SMPs. There might be an expected increase in the use of realignment as a strategic option – this may be particularly apparent as the "long term" will be up to 100 years hence, rather than the 50 years used in the first round of SMPs. This is supported by greater social inclusion in the process, which should allow greater consideration of the breadth of stakeholder interest.

Coastal defence strategies

A coastal defence strategy covers a smaller section of coast in more detail than an SMP and it describes how to achieve the adopted SMP policy. Strategies usually include detailed analysis of coastal processes, together with detailed analysis of economics and strategic environmental assessment. Under current guidance, they are expected to look at a period of 100 years from the present. In some cases, the additional information gathered may prompt revision of the SMP policy. Where managed realignment is the preferred policy, the strategy will investigate possible new defence alignments and the technical, economic and environmental implications of different options. However, final selection of an option is usually made at a subsequent stage, as part of a project appraisal report (PAR).

Coastal habitat management plans (CHaMPs)

CHaMPs are non-statutory plans aiming to assist in identifying the best strategic approach for coastal and estuarine habitats and their associated ecological interests where there is a conflict between coastal evolution, coastal defence measures and the needs of Natura 2000 sites (see also Section 3.2.5).

In a few locations, English Nature and the Environment Agency have developed coastal habitat management plans (CHaMPs). They are intended to provide a framework for the management of designated nature conservation sites of European interest in the context of existing Directives and Regulations. The seven pilot CHaMPs cover both coastal (eg Dungeness and Pett Levels) and estuarine (eg Essex coast and estuaries) locations and contain an understanding of the geomorphology of the study area. Further information about the CHaMPs and copies of the plans themselves can be obtained (www[3]).

Catchment flood management plans (CFMPs)

A CFMP is strategic document for the catchment-wide management of flood risk. It looks to a 50-year horizon, attempting to identify the measures required for successful and sustainable flood management within that timeframe. A CFMP would typically comprise four phases, namely inception, development of catchment (including tidal rivers/estuaries) understanding, development and appraisal of flood risk management policies and finally dissemination.

The Environment Agency and Defra are to develop CFMPs for river catchments in England and Wales in a similar way to SMPs covering the coast. Following the preparation of a draft set of guidelines for the plans, pilot studies were commissioned for five rivers, chosen to provide a representative sample of river catchments within England and Wales. Once the pilot CFMP studies have been completed, the plan guidelines will be finalised. Some scoping and data gathering exercises are being undertaken before the pilot studies and guidelines are finalised.

Biodiversity action plans

Biodiversity action plans (BAPs) originate from the International Convention for Biological Diversity signed in Rio de Janeiro in 1992. The UK BAP was prepared in 1994 and forms a blueprint for conserving the range of species and habitats in Britain, whether specifically protected by legislation or not. This was followed in 1999 by a Maritime Habitats and Species Biodiversity Action Plan volume that covers all coastal habitats and by local BAPs covering some counties. The national and local BAPs set specific targets for habitat creation to offset previous and predicted losses and provide ecological enhancements. Habitats for which targets have been set that are important in the coastal context include saltmarsh, sand dunes, vegetated shingle, mudflats, maritime cliff, grazing marsh and reedbeds. During 2002/03, English Nature held workshops to identify biodiversity opportunities in several maritime natural areas (www[5]).

Defra has incorporated the objective of meeting BAP targets into its high-level targets for flood and coastal defence. These require operating authorities to avoid damage to environmental interest, to ensure no net loss of BAP habitats and to seek opportunities for environmental enhancement.

BAP targets for saltmarsh, mudflat and other intertidal habitat creation are becoming an increasingly important driver for managed realignment in estuaries. For example, in the UK, it is estimated that 100 ha/year of saltmarsh needs to be created to account for future change, plus 40 ha/year over the next 15 years to replace past losses between 1992 and 1998 (www[4]). These targets are being divided up between natural areas (www[5]) and the identification of sites for habitat creation is being done through coastal strategy plans (see Coastal Strategies above). As much of the intertidal habitat that has been, or is predicted to be, lost is protected by the Habitats Regulations, government funding may be available for managed realignment schemes where the primary purpose is intertidal habitat creation (see Section 2.4.2).

River basin management plans

Going forward, river basin management plans will also be important insofar as they will provide a statutory basis for the *measures* necessary to achieve good ecological/ geomorphological status under the Water Framework Directive (see Section 1.2.2), and some of these measures might involve managed realignment.

1.4.2 Opportunistic managed realignment

Managed realignment may be termed "opportunistic" when it arises from "bottom-up" factors that relate only to a particular site. Almost half of the realignments to date might be considered in this category. Examples include:

- a landowner deciding to opt for realignment to create habitat
- an opportunity might arise to reduce defence maintenance costs, where inspection identifies a problem
- a landowner may be able to obtain Countryside Stewardship funding for a realignment scheme
- a conservation body may wish to create a new nature reserve.

Although opportunistic managed realignment is less likely to be constrained by lack of community support or lack of compensation, the same consents will need to be obtained as for strategic managed realignment. Difficulties may occur because of potential adverse effects on adjacent areas of coastal or estuarine dynamics; however, nature conservation bodies may not consider the geomorphological and hydrological effects of realignment to have an adverse effect on the integrity of intertidal and supra-tidal conservation features. Ideally, opportunistic managed realignment should take place within the strategic framework, enabling the strategic vision to be realised in the most acceptable way for all stakeholders. Notwithstanding this, opportunistic managed realignment is likely to continue to have a significant role in the near future.

1.4.3 Legislative requirements

A range of legislation may influence the development of future schemes. Managed realignment may result from the need to provide mitigation or compensation, under the Habitats Regulations, as a result of proposed development or coastal squeeze. There is a general presumption that the (re-)created habitat should be provided as close as practicable to the habitat that is lost to maintain the overall coherence of the Natura 2000 network of European sites. Although the need arises from legislation, the same consents needed for other schemes will still have to be obtained and impacts as a result of the realignment will need to be addressed. Thus a strategic view of the realignment is still important. Increasingly, the requirements of the Habitats Regulations are being recognised in the range of coastal plans and strategies. This enables the benefits of a strategic approach to be realised.

Another driver for managed realignment is the delivery of the Government's Public Service Agreement (PSA) target to have 95 per cent of all SSSIs in favourable condition by 2010. English Nature has published a report to help in this assessment (English Nature, 2004).

Re-establishing the ecological integrity and ecological function, as stipulated within the Water Framework Directive, may also prove to be key drivers for managed realignment (see Section 1.2.2).

1.5 STAKEHOLDER PARTICIPATION

1.5.1 Introduction to stakeholder participation

Effective engagement with local communities and other stakeholders is key to successful implementation of managed realignment schemes. Participation can help to:

- understand legitimate concerns and interests so that these can be incorporated into scheme design
- explain and convince the local community of the merits of a scheme
- manage expectations of what can be achieved
- develop stakeholder ownership.

Although stakeholder involvement is not specifically a design issue, it is of sufficient importance to ensuring effective implementation that it should be considered throughout the planning and design process.

Detailed liaison with stakeholders on a wide range of issues is time-consuming, costly in management time and may lead to outcomes that are not focused on the primary aim of the project. However, experience suggests that, in some instances, it may help avoid greater delays and hence expense. Some proposed managed realignment schemes have even been abandoned following sustained public opposition, demonstrating that trying to impose a solution on a neighbourhood may not be appropriate.

Persuading local community members to embrace the scheme may be difficult, especially if a defence has a long history or if locals have invested time and money in certain land practices. People are often emotionally tied to land and property, and change can be hard to accept. Local communities may also have very little knowledge of managed realignment and may not understand why defences cannot be retained.

1.5.2 Involvement throughout the design process

Raising the ideas behind managed realignment at an early stage provides time for people to absorb the concept and formulate questions and contributions to the design process. Where stakeholders have a good understanding of the issues they are more likely to accept the scheme. Stakeholder involvement and communication with local groups should be seen as valuable opportunities to educate the public about the benefits and constraints presented by managed realignment (and flood management generally), and to be able to respond to rumours, assumptions and questions.

Participation methods

Some schemes found that this need could be addressed effectively through a steering group or forum that included elected representatives of local communities as well as statutory consultees and interest groups. Where it can be justified by the proposal's size and economics, the use of professional facilitators with specialist communication skills is recommended and should be included in the plans and budget. Stakeholder participation activities may include:

- public meetings

- private consultation in peoples' own environment (eg at private houses)

- information newsletters/bulletins

- local television and radio programmes and newspaper reports

- study visits to other managed realignment sites.

Addressing scheme issues

The types of issue that arise and can be resolved through this process include:

- routeing of construction traffic

- protecting natural resources such as vegetation, fish and wildfowl used under common rights

- creating new intertidal or freshwater habitats, such as lagoons or reedbeds in borrow pits

- acquisition of land within the realignment area in a way that is perceived by local communities as fair. Ensuring or facilitating the construction of counter walls to protect adjacent land uses

- potential economic benefits such as job opportunities in natural area management and visitor services.

During evaluation of project options, available data and information should be collated and reviewed. This process may identify options that are suitable from a design perspective (for achieving the scheme aims and objectives) while also maximising opportunities with stakeholders and overcoming constraints. Conversely, some design options may need to be screened out of the process where the stakeholder issues are considered likely to be too significantly restraining. The information collated at this stage can also be used to help define success and failure criteria for the scheme and can feed into later stakeholder involvement activities.

Box 1.2   *Stakeholder engagement strategy*

The Environment Agency handbook for scoping projects (Appendix C) and draft procedural guidance for SMP2 usefully summarise public consultation methods and their application.

Times, methods and procedures for stakeholder engagement need to be planned before work starts on design. A "stakeholder engagement strategy" is a key tool. It sets out both the vision for stakeholder engagement, the stakeholders to be involved, the methods employed and the responsibility for implementation at each stage of the design process. Establishing the vision requires balancing the needs for stakeholder inclusion and conflict minimisation with the willingness to allow others to help shape the process and to resource implications. The strategy should set out decision-making and administrative responsibilities, opportunities for representation and mechanisms to resolve differences of view; once agreed, these activities must be managed.

The principles influencing a stakeholder engagement strategy include the following.

1. *Inclusivity*. The initiation of the design process should indicate whether a participatory or a consultative approach is adopted and outline the extent of wider community involvement.

2. *Transparency*. Timely, accurate, comprehensive and accessible recording of representations, decisions and their justification is required to track decisions. The strategy should indicate who has responsibility for this.

3. *Appropriateness*. The range of stakeholders, their level of involvement and likely knowledge, the potential for differences of view and the opportunity to raise awareness will influence the approach adopted.

4. *Ownership*. The strategy should place emphasis on the need to develop ownership of the scheme by the stakeholders and to manage their expectations throughout the process.

5. *Clarity*. The roles of different "players", including where final decision-making lies, must be made clear in the strategy.

6. *Comprehensiveness*. The strategy should cover all stages, including plan dissemination and arrangements for reporting on stakeholder engagement.

Those responsible for preparation must decide whether a participatory or a consultative approach is appropriate. A participatory approach implies that other bodies contribute to the decision-making process, including, eg, English Nature, planners and/or community interests. Even if a participatory approach is adopted, wider consultation will still be needed. This involves seeking third party views and comments and considering their representations. It may, or may not, result in information, ideas or design being amended. The consultative approach leaves decision-making with the stakeholders or their representatives, but "consultative partnerships" can be developed, bringing individuals and groups together to provide advice through advisory groups or committees. In selecting the people and organisations to engage in the process, it is important to remember the need for inclusivity and the context of the plan in question. Local knowledge should be used to identify people and groups who are likely to be affected by the plan.

1.6   SUCCESS AND FAILURE CRITERIA

Success criteria (see examples in Table 1.1) should relate primarily to the purpose of the scheme. Each project will have different aims and therefore different measures of success. Interest groups are also likely to have varying interpretations of what constitutes "success", and it is important to establish clear, agreed parameters at the outset. It may not be possible to achieve all the objectives identified initially. Measures of success need to be quantifiable but not so prescriptive as to constrain the outcome. It is also useful to set out what would constitute failure, so that the envelope of acceptable outcome is defined. This helps in post-scheme monitoring, and also in explaining to stakeholders what the project will deliver.

Secondary criteria may also be defined in terms of fulfilling opportunities and managing constraints, but they should not be confused with the main purpose. Such criteria may be identified either by scheme promoters or through stakeholder participation (see Section 1.5). They may be considered as aspirations that are advantageous, but not fundamental to the success or failure of the project. If the primary reason for a managed realignment were to create saltmarsh habitat, secondary criteria might be increased public access, creating bird-roosting areas (Atkinson *et al*, 2001), providing bird hides, or navigation of the new area. Each of these secondary criteria has potential to compromise the primary purpose of the scheme, but it still may be possible to accommodate them in the overall design of the project.

While success criteria for flood defence standards are often relatively easy to define, criteria relating to habitat creation are likely to be more complex. Lack of adequate ecological knowledge often hampers the defining of habitat criteria, which may only be fulfilled over a long timescale. Government conservation advisers may find it acceptable to have a degree of uncertainty in the specific species or habitat that a managed realignment project may bring about. "Functional" success may be defined as the ability of ecosystems to support food chains, attenuate storm action and improve water quality while adapting to natural changes. "Compliance" success reflects the need to meet the requirements of legislation or an agreement – for example, habitat should ideally be "like for like", of at least equal area, in an adjacent location, and ideally established before the habitat to be replaced is lost. Some approaches for evaluating habitat restoration projects and defining success criteria are discussed by Kentula (2000) and Short *et al* (2000).

In practice, success criteria for ecological objectives need to be kept simple and expressed in measurable terms, such as areas of mudflats and saltmarshes or numbers of birds feeding and/or roosting on site. It is to be expected that sites will change over time and for there to be uncertainties about how they will evolve, but it is often better to take opportunities offered by managed realignment even when there is a degree of risk (English Nature, pers comm, 2003).

An ongoing Defra/Environment Agency research project, "Suitability Criteria for Habitat Creation" (Project FD1917) aims to produce an electronic decision tree for users to assess the potential of specific sites for habitat restoration schemes. This encompasses knowledge on the criteria for growth of natural saltmarsh habitats and on the selection of sites for habitat restoration. Another Defra/Environment Agency research project (FD1918) is also providing guidance on monitoring and habitat quality that can be used to assess the success of a realignment site. Section 3.2 provides details of other research projects relevant to target-setting.

Failure criteria may particularly relate to the management of avoiding potential adverse impacts or overcoming perceived constraints. If the management or mitigation is not successful and the adverse impact is realised, then the scheme may be judged to have failed even if some success criteria have been met. To assess success or failure requires a well-designed monitoring programme. Funding for monitoring that quantifies change is not always justified, and qualitative assessment may need to be used.

The timescale over which success and failure criteria are to be evaluated needs to be determined at the outset. It is likely that some successes will only come to fruition some years after the inundation of the site. Realistically, a time horizon of five to 15 years is appropriate for monitoring the effects of a scheme and informing future decision-making. When considering the sustainability of both flood defences and habitat creation, planning needs to encompass much longer periods, of 50–100 years.

Table 1.1 *Examples of success and failure criteria that have been used for managed realignment schemes with different aims*

Aim	Success	Failure
Realignment of shingle ridge for flood defence	Realignment inland by 200 m of 1.5 km in length	No realignment
	Maintaining 1:20-year standard of flood defence	Less than 1:10-year standard of protection
	Shingle vegetation established in five years	No vegetation established
	Reduced annual maintenance costs of £50 000 over next five years	Increase in maintenance costs
	Roll-over of ridge under natural processes	No ridge roll-over; loss of shingle under catastrophic failure
	Protection to a road and property at 1:50-year standard	Protection not given to property at 1:50-year standard
Habitat creation in estuarine creek	Realignment of 45 ha of land contributing to the functioning system	No realignment
	Creation of new saltmarsh habitat	Net loss of intertidal habitat inside/outside site
	Site used as bird-roosting/feeding area	No bird roosting or feeding
	Vegetation established in 10 years	No vegetation established
	Reduction in flood defence costs over the next five years of £50 000	Increase in flood defence costs
	The mudflats are suitable for shellfish	Damage to shellfishing outside site
	Footpath maintained landward of site for public use	Footpath compromised
	Site adopted and managed by local wildlife trust for 10-year period	Site unmanaged
	Contribution to costs from more than one organisation	Insufficient funding to proceed
Estuarine site for development mitigation or compensation	To ensure that a development's legal obligations under the EC Birds Directive and Habitats Directives are met	Failure to meet obligations resulting in court action, adverse publicity or impact on company share price
	Realignment to create at least 50 ha of mudflat and 50 ha of saltmarsh	Site fails to create habitat
	To achieve a functioning system within 15 years and hence provide a replacement environment of equal conservation value to that lost through the development	Net change of inter tidal habitat (saltmarsh and mudflat) inside/outside site not providing a functioning system with the existing estuary
	Site sustainable over a 50-year period	Site unable to adapt and respond to sea level rise and becomes either *terra firma* or sub-tidal
	Where possible, the scheme should be consistent with the requirements of other strategic plans and not compromise strategic options identified within them (including shoreline management plans (SMP), flood management strategies and coastal habitat management plans (CHaMPS))	The site selection and evolution compromises the ability to carry through already defined strategic objectives
	Construction to budget in a period of five years (by 2008)	Increase in costs above budget and delivery after five years
	Scheme design acceptable to English Nature, Environment Agency and RSPB based on scheme impacts and targeted habitat creation	Scheme design not acceptable to English Nature, Environment Agency and RSPB
	Archaeological interests accounted for and recorded	Damage to archaeological interest without recording
	Length of footpath access maintained or enhanced	Footpath lost or future use compromised

1.6.1 Multiple objectives

In practice, most realignment schemes have multiple objectives, which may include both the underlying purpose(s) and opportunities. Most frequently, these will include sustainable flood defence and intertidal habitat creation. Whatever the original driver for realignment, the scheme design should address as many beneficial objectives as feasible. This will usually result in benefits such as:

- support for the project from potential objectors
- an improved priority score under flood defence prioritisation system
- an improved chance of securing technical support and/or funding from a variety of sources
- the ability to meet a wide range of societal needs.

In adopting multiple objectives, it is important that the original aims are not compromised to accommodate the other objectives. Stakeholders may have differing aims or objectives and may change their view as plans proceed. For example, delivery of biodiversity enhancements may be important to some stakeholders but not to others. At a strategic level it may be possible to meet the needs of all stakeholders by co-ordinating them across several schemes.

Identifying opportunities and working with multiple stakeholders can be time-consuming, but it can deliver advantages. Stakeholders may have different resources to commit to the project and different knowledge, skills and experience to contribute. In addition, the potential benefits of identifying opportunities and facilitating their delivery include shared costs, shared risks, shared ownership, reduced objections and a more holistic outcome.

Using multiple objectives

The general lack of familiarity with managed realignment makes it important to promote the range of benefits and technical requirements. Promoting the benefits of a managed realignment site using multiple objectives can help persuade different stakeholders to accept a project, including the local public who usually have diverse views and interests.

There is a need to explain to stakeholders that managed realignment schemes can offer a package of measures able to meet multiple objectives. This may include aspects such as fishing and/or fishing nursery areas, pollutant treatment, a "green" environment, sailing, flood defence improvement, reducing the burden of costs on taxpayers, producing an environment for future generations to enjoy, shooting, niche agriculture and so on. Experience suggests it is not advisable to approach the public with only one objective or driver for a managed realignment site – all the potential benefits should be communicated.

Box 1.3 *Example of challenges for a "single issue"*

> At one managed realignment scheme, the public was informed that realignment was being undertaken for habitat creation. This was seen as flooding peoples' property "...for a pair of redshanks", creating great antipathy to the scheme. After the perception had been generated, it was difficult and time-consuming to get people to consider other views objectively.

Where only one objective or driver is put forward as the reason for realignment, it may not meet general acceptance. The perceived lack of success may then have a detrimental effect on the implementation of other managed realignment schemes. Putting multiple objectives in place reduces the risks of the project failing, because if one of the objectives does not meet the criteria but the others do it can still broadly be perceived as a success. It can also make it more acceptable to stakeholders and local community members. It may be possible to flag up failed objectives as important issues for other local realignment opportunities, which should minimise objections and make it more likely that the needs of all stakeholders can be met over a range of projects. Such linkages should be publicised once achieved.

1.6.2 Communicating the lessons learnt

As the managed realignment scheme develops, periodic reports should be issued in an easily understood and accessible format. They should be budgeted for in the project planning stage. The frequency of reporting should be in proportion to the size, complexity and sensitivity of the project and the extent to which the scheme has developed. For large or sensitive projects, quite frequent reports may be required at inception, perhaps annually for the first five years and subsequently at significant milestones. The timing of reporting might also be aligned to monitoring requirements and the provision of updated or new information about the project.

The reporting should include the problems faced and how they were overcome. It is advisable to be open and transparent about what has occurred, including detrimental outcomes or where the reality of what occurred differed from what was expected, both during and after the scheme implementation. This approach is more likely to reduce problems with the local community in future years, especially if part of the scheme develops differently from the project expectations and changes have to be made. Also, understanding the issues will help professionals to develop further tools and techniques to support managed realignment and to share best practice solutions to problems. This will help reduce uncertainty and risk in future schemes and would provide a positive contribution to future R&D work.

Particularly for large or sensitive sites (or several smaller linked sites) public perception issues should be addressed through local or national media (see Section 1.5). For small or non-contentious projects, it may be sufficient to use local media, such as community, church or parish council newsletters. Often the greatest problems arise from apprehensive residents, so these and other local communication channels should be used to keep people informed as the project progresses. Local television and newspapers are important additional media that can deliver messages regionally, while for large schemes with high risks it may be helpful to include national television and newspapers. It is important to ensure that messages come from an identified individual (a spokesperson) for consistency and familiarity. Experience has shown that education needs to be a continuing process if managed realignment is to attain the same level of acceptability, in the public eye, as other existing flood management options.

In summary, to increase the uptake of managed realignment as a viable and effective option, there is a need to create "good sites" and to distribute information about them to the public, the media and design professionals.

Part I – key conclusions

WHY MANAGED REALIGNMENT?

Managed realignment may be undertaken for a variety of purposes including flood management, habitat creation, or in association with new development. The reasons for managed realignment as well as the characteristics of the particular location will help determine the preferred approach.

1   Managed realignment may have one or more drivers. This provides an opportunity to meet a variety of stakeholder needs, but requires careful management to ensure the drivers have complementary objectives.

2   The need for managed realignment may be impelled by strategic management plan objectives, *ad hoc* opportunities or legislation.

3   Communities may oppose managed realignment for reasons that include emotional ties to assets, economic loss for the individual and change to rights of access (particularly footpaths). These issues need to be addressed through public involvement and/or participation from an early stage and for some years following the scheme implementation.

PART II

How and where managed realignment can be achieved

> Part II discusses whether managed realignment is appropriate (once it has been selected as an option) for a particular site and explains how it may be delivered. This section is intended for a wider audience; including coastal managers, geomorphologists, consenting authorities and environmental advisers. It explains the decision process that leads to a preferred approach.

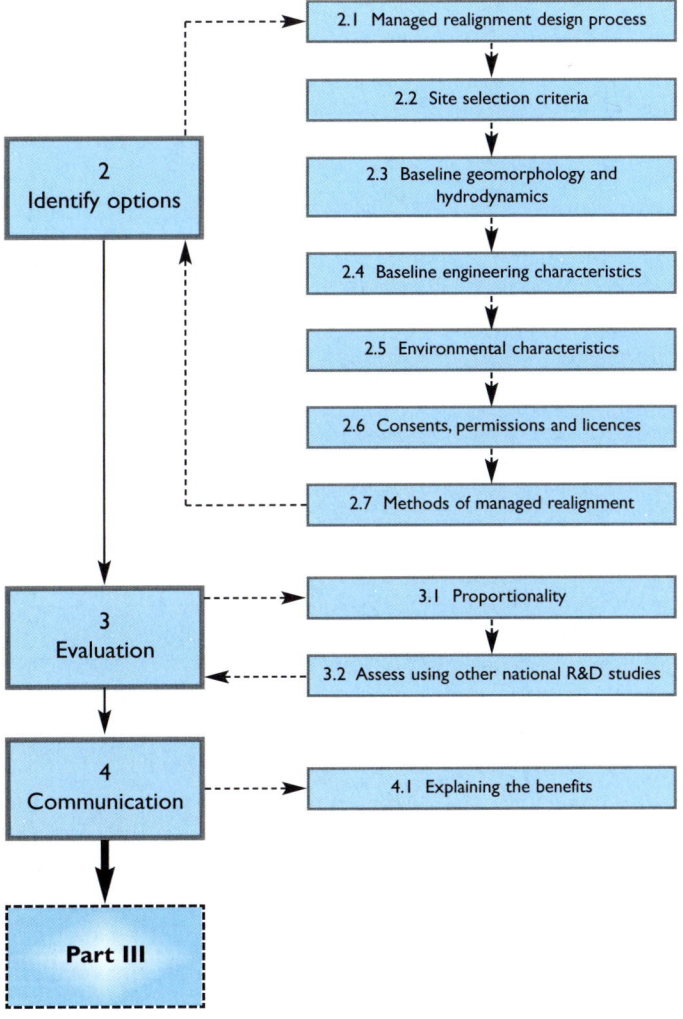

2 Identifying options for managed realignment

2.1 INTRODUCTION TO THE PLANNING AND DESIGN PROCESS

The design of a managed realignment scheme must be considered as a dynamic process that involves a number of stages, some of which must be iterative in order to maximise the effectiveness of a scheme. Deciding whether the site is appropriate, and the subsequent design, is dependent on the following factors:

- the specific aims and objectives of the scheme (ie what it is trying to achieve)
- the identification and influence of opportunities and constraints (ie what can and cannot be done)
- the range of available implementation options that exist
- the technical feasibility and economic viability of those options in the specific context of the scheme under consideration
- the environmental and social acceptability of potential within-site changes and wider-scale impacts of the scheme
- the risks and uncertainties that exist.

The managed realignment design process is shown in Figure 2.1 (overleaf). The specific approach to managed realignment is defined after baseline assessment methods have been applied.

While the use of managed realignment in managing the coast is not new, neither is the approach to managed realignment a long-established or familiar process. Time invested in the planning and executing of a project will help ensure success (Dixon and Wright, 1997; Dixon *et al*, 1998). Conversely, trying to force shorter timescales, or failing to identify key milestones in the project, may lead to disappointment and extend project delivery. A critical path for the entire process is needed, but separate critical paths on consents and licences are also required.

The project plan must define all risks and address how the risks will be managed or designed out; it is unacceptable to have no course of action to address a risk. For example, if the site size is an issue and the site is located at the top of an estuary (where impacts will be greater) a similar-sized site at the mouth of an estuary or on the open coast may provide an acceptable alternative.

If financially limited, the money available may constrain the approach to design and what may be achieved at the site. Nonetheless appropriate hydrodynamic and geomorphological assessment must be undertaken, as a minimum, to manage risks and impacts on the environment. If finance is less constrained, innovative approaches may be possible for construction and potential future site modification.

PART II

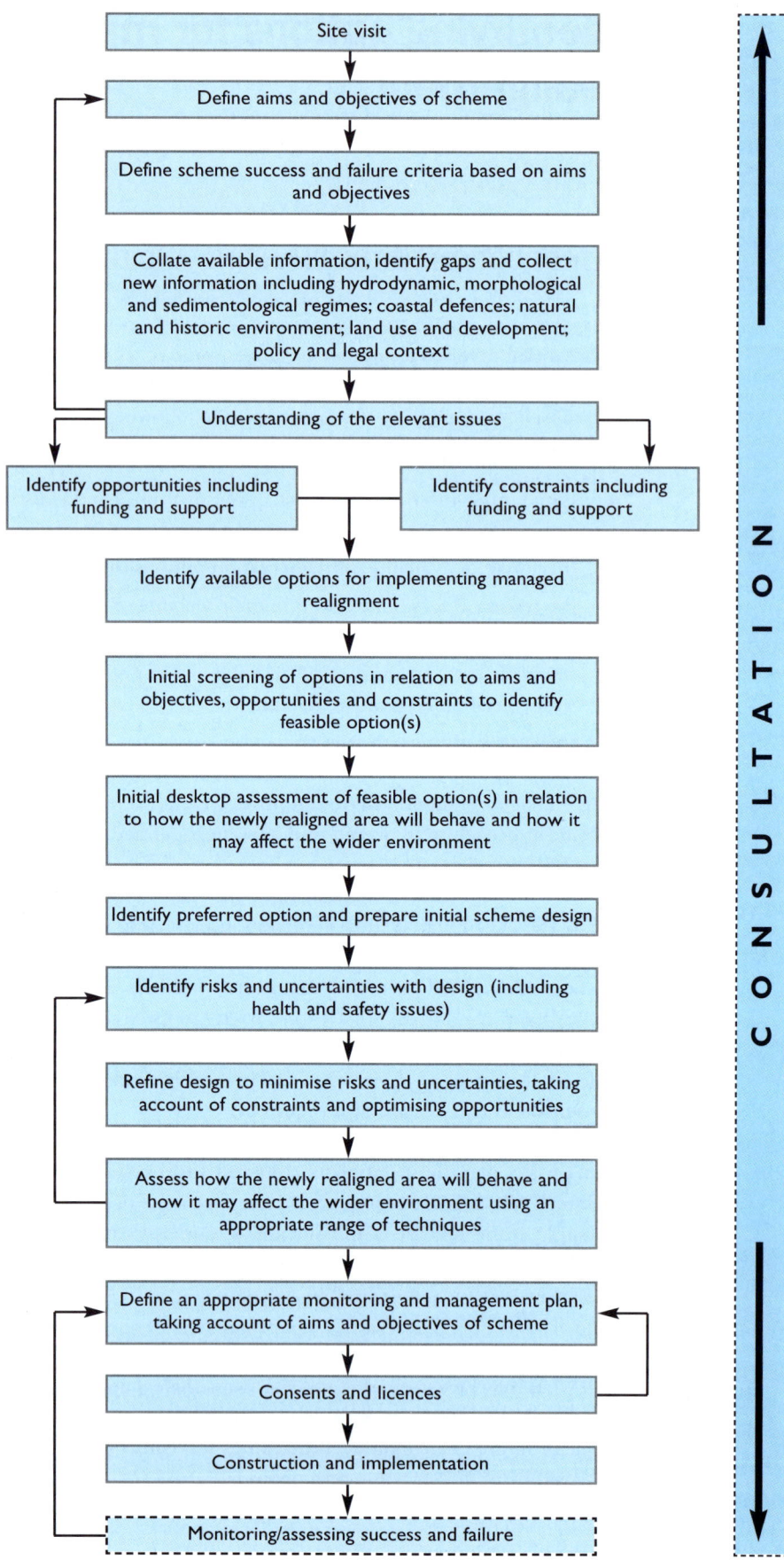

Figure 2.1   *Managed realignment: planning and design process. The starting premise of this diagram is that managed realignment has been determined as the strategic approach*

In order to assist with the identification of the aims, objectives, opportunities and constraints for a particular scheme, it will be necessary to involve stakeholders in consultation at an early stage. This process should continue throughout the project in order to identify and accommodate relevant issues and interests (see Section 1.5).

2.2 SITE SELECTION CRITERIA

With many competing factors relating to site selection, it is necessary to define priorities and establish a way of comparing the suitability of different sites.

Where flood defence is the primary driver, selection of a suitable site should be compatible with coastal or estuarine processes but will mainly arise through economics, ie where managed realignment is a more economic whole-life solution than holding the line or no active intervention. Factors contributing to site selection might include:

- a high level of confidence that existing estuary or coast defence is unsustainable at that location and managed realignment offers a potential solution (a strategic view)

- economic viability for the existing defence is low: a benefit/cost ratio of less than 1 strongly favours consideration of realignment

- the residual life of existing flood defences is low, poor defence condition being preferable for realignment

- it is economically preferable to realign to naturally occurring high ground, as a shoreline would be created and a new, engineered defence would not be required. This may be easily achieved where the floodplain is narrow

- shortening of the length, or reduction in the size, of existing defence is preferable

- there is limited impact of managed realignment on erosion outside the realigned area and low wave exposure for the site (quiescent sites may be preferable)

- the existing site has suitable elevation, particularly where habitat creation is an objective. For example, a site level above the approximate MHWN with a gradient of 1–2 per cent might be beneficial where saltmarsh habitat is desired but a lower site elevation may be required for mudflats. This reduces the need for earthworks

- there is low potential impact on local environment, eg absence of freshwater or terrestrial habitat within SAC, SPA or SSSI; scheduled monument, archaeological sites, public footpath, forestry etc

- there is an absence of any built development within or close to the proposed realignment site. In some cases, local defences or ring banks may overcome the problem that would otherwise be caused by a small area of built development in a realignment site

- there is land available. This is an important consideration because land ownership boundaries and the degree of interest and participation by individual landowners may constrain the site boundaries

- the area has lower agricultural land classification: Grades 3b, 4, 5 are preferable

- an absence of contaminated land, including landfill sites and refuse-filled seawalls

- the site has ease of access and working within the site

- there is a potential source of funding and/or financial support.

PART II

Where the principal driver is habitat creation (whether to meet BAP targets or as compensatory habitat in respect of a development), the site's location within the wider environment will be important, as will the physical conditions determining the type of habitat that might develop. It is likely that a range of sites – for example within one or more estuaries – may be considered to meet the requirement for habitat creation. The generic site selection review as part of Defra/Environment Agency project FD1917 can assist in this process (see Section 3.2.1). It should be acknowledged that the "ideal" site to re-create the lost habitat might not be immediately adjacent to the area of impact and where possible should be selected to be self-sustaining.

A possible approach to site selection is through a hierarchy of criteria determined on a site-by-site basis; this avoids the assignment of arbitrary weightings for different factors, by grouping them into categories based on the criteria set out as required/desired. Sites are then scored against each criterion and the preferred sites are those that meet most of the key criteria. Then within those, the sites that meet most secondary and tertiary criteria can be determined.

Box 2.1   *Example of an unsustainable approach*

A managed realignment site had to create mudflat as mitigation for development under the Habitat Regulations. The site selected was only slightly larger than the area required for mitigation. After realignment, the site was found to be too high and had to be lowered to enable mudflats to develop. It is now accreting sediment to the point that saltmarsh may form. Thus further intervention is needed to reassure stakeholders that mudflat will be maintained, to meet the original, legal, requirement. This will either require redesigning the site, continued maintenance to avoid accretion, or an alternative sustainable site will have to be created that will maintain mudflat. Thus further costs will be incurred.

The mitigation could have been addressed using a larger site that would have a mix of salt-marsh and mudflat. Parts of the larger site would form other habitats, but it may be necessary to have a larger site to be able to deliver the area of mudflats required. Alternatively a more exposed and low-lying site might be designed to provide sustainable mudflat.

2.3   BASELINE GEOMORPHOLOGY AND HYDRODYNAMICS

Relevant baseline information relating to coastal or estuarine processes and morphology is necessary in order to:

- inform the initial identification of sites that are potentially suitable for managed realignment.

- assist in the design of a managed realignment scheme.

- assist in the assessment of impacts of a managed realignment scheme on the wider environment.

A proportional approach to collation of information should be adopted. In most cases, the minimum baseline information required will include the following.

1   Topography of the managed realignment site and adjacent hinterland (from these data, parameters such as site area, gradient and level can be determined). It is important to understand the relationship of levels to landward and seaward of the defence line. The available approaches may be divided into land-based and aerial surveys, which are outlined in Appendix 1

2   Tidal levels and tidal range in the vicinity of the managed realignment site (this, combined with the topographic information, can be used to determine the extent and frequency of tidal inundation and tidal prism). Actual tide levels may vary significantly from those predicted astronomically or listed in tide tables.

3   Wave activity affecting the site (to help determine erosion and sediment transport both outside and within the site after realignment; see French *et al*, 2000).

4   Geology and sediment composition of the managed realignment site and the surrounding environment (this has implications for erosion within the site and in adjacent areas, the supply of sediments to the site to enable vertical accretion, and the geotechnical properties of the within-site "basement". The characteristics of the basement are relevant, as any new defence will be constructed on them or natural features such as gravel barriers or sand spits may migrate landwards across them).

5   In estuaries: the tidal prism, bathymetric representation and sediment budget of the estuary. This enables post-scheme changes to be compared against a pre-existing baseline condition and to identify potential wider benefits to the functioning of the estuary.

6   In coastal systems: a conceptual understanding of the littoral sediment transport processes, eg sediment budgets, and how they may be altered by the managed realignment scheme (including the site tidal prism). This should consider both longshore drift and cross-shore changes to bathymetry (hence altering the wave climate locally) and current rates of accretion and erosion.

It is important to ensure the level of data collection is proportionate to the size and complexity of the site and the potential impacts of the managed realignment. All but the smallest or most straightforward sites will need some amount of survey. This might include human site use, landscape assessment and flora and fauna, in addition to the physical issues listed above.

The above information can be collected using a variety of techniques to varying levels of spatial resolution, but much is often available from existing studies or information sources, ranging from national research and development studies, through regional coastal plans, to local site-specific measurements. Many schemes will have much of the information needed in the relevant shoreline management plan, coastal strategy, coastal habitat management plan or other strategic study (see Section 3.2). Some examples of wider, useful, information sources include:

- British Geological Survey maps (solid geology, drift geology and seabed sediments)
- the countryside agencies' geological/geomorphological conservation review (GCR) sites have separate reports on the site interest and management with information linked under the coastal geomorphology block of sites
- Ordnance Survey 1:10 000-scale maps
- Ordnance Survey first edition (and subsequent) maps, which give an indication of the extent of past anthropogenic impacts, eg reclamation, and natural change in the coastal or estuarine system. Although significant change will pre-date the first edition, this can provide useful information for the past 150 years or so
- Admiralty tide tables (tidal levels, range and asymmetry) may be used for preliminary ("first order") assessments, but should be interpreted with caution and, where possible, supplemented with other data or field measurements to verify their applicability at the realignment location
- Admiralty and port authority charts (for identifying the presence of bathymetric features such as sand banks and channels, and enabling first-order estimates of the magnitude of estuarine tidal prisms)
- Futurecoast (Halcrow Group, 2002a) for a range of coastal, and some estuarine, data
- Environment Agency digital terrain modelling and regional survey programmes (including LIDAR and photogrammetry).

It is recommended that, before collecting scheme-specific data, information from available sources is collated and evaluated for its potential contribution to the baseline understanding. While it is acknowledged that not all parameters will be available to a desirable level of accuracy or spatial resolution many parameters will be identified through this approach.

2.4 BASELINE ENGINEERING CHARACTERISTICS

2.4.1 Performance of existing defence

Where flood defence is the main driver for a managed realignment scheme, an assessment of the existing standard of protection against breaching and overtopping will be required to determine whether managed realignment is an economically viable alternative to other approaches (HR Wallingford, forthcoming). Such an appraisal is also valuable where drivers other than flood defence are promoting the realignment.

Although not a conventional measure of defence performance, it is important to consider the degree to which the existence of a defence has caused a variation in elevation between its landward and seaward sides. If the crest level of the defence is high enough it will prevent the normal processes of sedimentation and erosion occurring to landward; this can lead to a difference in level compared with the seaward side of the defence. Separation from the physical processes might lead to a relatively higher landward level in an erosive environment and a relatively lower level in an accreting environment. Such patterns may be complicated where the seaward environment is generally eroding but saltmarshes or dunes are increasing in elevation immediately in front of the defence, or where reclamation of land has altered the seaward environment, encouraging greater sedimentation or erosion.

Factors that determine the degree of difference include the extent of changes naturally occurring in the seaward environment, the processes of deflation (shrinkage) or consolidation of the landward environment, and the time over which the land has been separated from the natural processes. In some agricultural practices low bunds are used, allowing regular flooding of the land. In these cases the landward side is less divorced from the natural processes, but the bunds may encourage artificially high accretion to the landward side. In general the smallest difference compared with the natural environment will be found where defences are low, land is frequently flooded with associated sedimentation, and/or has been separated from the coastal and estuarine process for the shortest period of time. The variation between seaward and landward levels can be a significant design issue.

Many existing flood defence structures were constructed many decades ago (and may even be of archaeological interest), before the development of modern design methods, so information describing design methods and standards is often not available. It is important to bear in mind that defences may not be totally removed under managed realignment and consequently may decay over time (HR Wallingford, forthcoming) and alter the realignment area in the future.

Figure 2.2   *Poor defence conditions may lead to adopting managed realignment. How such defences may decay over time needs to be understood as part of the design process*

2.4.2   **Contaminated land**

The presence of potentially contaminated materials should be investigated at an early stage in order to identify any constraints that may affect the design process. A desk-based study of the area, utilising historical Ordnance Survey mapping and local historical resources, should be undertaken before site investigation to provide the scope for any future detailed investigation. Intrusive site investigation may be required and will generally involve trial pits and boreholes to characterise the sub-surface and provide samples for further analysis. Once the nature and extent of any contamination has been determined, a risk assessment will be undertaken to establish the significance and to develop a remediation strategy. Risk assessment involves examination of the source, pathways and receptors; a risk exists only when all three of these are present. Further information can be found in *Contaminated land risk assessment: a guide to good practice* (Rudland *et al*, 2001 and www[13]).

Managed realignment often requires the removal of, or excavation into, existing sea defences, which may expose site visitors and the wider environment to any potentially contaminated materials that were used historically to construct the defence or that were protected by it. Experience has shown that seawalls have been constructed with waste materials derived from local industries such as power station fly ash, clinker from foundries and gasworks or other potentially contaminated materials including general refuse. The use of these waste materials was not generally considered to present a risk to human health or the environment in the 19th and early to mid-20th centuries when many sea defences were constructed. Where land close to the shore is low-lying, it may have been raised by landfill in the past to prevent flooding and to allow beneficial uses such as grazing. These landfills may contain a wide variety of wastes, ranging from inert construction and demolition materials to hazardous industrial and household waste.

Another potential source of contaminants is agrochemicals, which may be mobilised or released to the marine environment following realignment. Landowners may have information on the use of these. In some cases, an ecotoxicology study should be commissioned before the project goes ahead. The pollution threat posed by agrochemicals can often be mitigated by a period of non-use before realignment to allow the contaminants to diffuse into the environment or to allow some compounds to break down with exposure to UV light (Imperial College, 1992, and Meakins *et al*, 1995). An alternative option may be to remove topsoil for use on other agricultural land, although this will lower site elevations.

The presence of contaminated land, either within the sea defences or protected by them, can affect a realignment scheme. Contaminants may cause pollution of surface and groundwater, pose unacceptable risks to humans as well as fauna and flora (with a particular identified risk for shellfisheries) and may require disposal to properly licensed facilities. Where disposal is necessary, the additional costs may make the scheme unacceptable. Where contamination is present within or adjacent to the sea defence, any material remaining after the realignment must be adequately protected from erosion, or removed.

Remediation options may involve treating the contamination, removing it from site or encapsulation. The precise method will depend on the nature of the contaminant and the nature of the realignment proposals. The cost implications of removing contaminated material to offsite disposal can be significant, typically in the range £50–150 per m^3 (Davis, Langdon and Everest, 2002), depending on the nature of the material. Added to this are costs relating to increased construction time and the impact on the local environment from transporting material to a new waste site. Any remediation should always be evaluated at an early stage to allow modifications to the scheme design to minimise the disturbance and to allow sufficient time for the necessary permits and permissions to be obtained from the regulatory authorities.

2.4.3 Geotechnical assessments

Tidal flows, wind and wave action can cause sediments of different sizes and/or characteristics to be eroded, entrained, transported and deposited. Methods of measuring sediment size include wet/dry sieving, settling and laser diffraction. It is important to consider the shear strength of the sediment and hence how easily eroded it will be (either *per se* or over time) as a result of tidal inundation. Vane testing, shear boxes, or direct erosion testing using flumes or wind tunnels (for wind-blown sand transport), set up in the field, can all provide insight to how mobile a sediment type might be. Assessing sediment properties in this way may help in determining the present erosion and sedimentation forces in the locality of the site and how these might be changed by the realignment. This might also have relevance on how the realignment will behave over time and the rate of change that might be expected; for example, strongly consolidated clays might not erode readily whilst unconsolidated peat or sands might be highly mobile. Such characteristics might influence where, when, and how creeks form in the site, and if so, how rapidly, thus providing feedback to how the site will perform hydrodynamically.

A range of geotechnical assessments of existing defences (HR Wallingford, forthcoming) may be required in the course of appraising and developing a managed realignment scheme, including:

- assessment of the geotechnical stability of existing flood defences and/or underlying strata

- design of geotechnical aspects of improvements to existing defences, for example raising and widening of embankments, including considerations of building complexity, settlement and bearing capacity failure

- assessment of the suitability of existing ground conditions for constructing new alignment of defence. The load-bearing capacity or likely consolidation of the sediment, which are both significant when dealing with muddy sediments or soft surfaces, will need assessing where a new defence is to be constructed or a dune/shingle ridge is expected to migrate to the area. Current penetration test (CPT) or odometer tests can be used in this respect

- geotechnical design of the new flood defences or closing banks, once the alignment has been selected, including considerations of building complexity.

Broadly, any such assessments will tend to follow a staged approach to maximise the benefit of existing information and site inspections before carrying out more expensive ground investigations (HR Wallingford, forthcoming; see also Section 5.5.2).

2.4.4 Site access

Access to the site may affect on the initial studies as well as the construction stage of the managed realignment. The following items should be considered:

- restrictions (width, height, weight and noise) for vehicles and plant on public highways leading to the site
- restrictions (width, height, weight and noise) for vehicles and plant on private roads leading to and around the site
- restrictions (draught, tidal level and cycle, navigation) on access by water for vehicles and plant
- the need for both statutory permissions and landowner/tenant consent to achieve suitable access
- the need for enlarged accesses and working areas for plant, given that many flood defences were constructed with greater reliance on manual labour than would be expected today
- the need to provide temporary haul routes
- the load-bearing of access routes and the need to undertake strengthening measures
- the restoration of access routes at the end of the project
- seasonal restrictions (either engineering or environmental).

A survey of site access should be carried out, including surrounding land, to identify how the works can best fit in with other activities. Any operations on site should follow good environmental practice (Petts, 1995; Millard and Sayers, 2000; Coventry and Woolveridge, 1999; Leggett and Holliday, 2002). The above factors should feed into considerations on the economic viability of a potential scheme (see Section 2.6).

2.5 ENVIRONMENTAL CHARACTERISTICS AND ASSESSMENT

Environmental characteristics of a potential managed realignment site include:

- the existing environment (environmental assets that may be lost, changed or disturbed)
- the potential environment (environmental assets that may be created or changed).

How these are defined depends to some extent on the objectives of the proposed realignment, particularly in respect of any habitats that the scheme intends to create. An introduction to the relevance of environmental impact assessment (EIA) and the types of assets that should be considered are outlined in the following subsections.

PART II

2.5.1   Environmental impact assessment

Environmental impact assessment (EIA) plays in important role in evaluating scheme designs and hence informing the design, implementation and monitoring processes, often setting constraints that the design will have to accommodate. The design of a scheme and the EIA process need to be conducted iteratively, therefore. The characteristics of a site need to be assessed and the managed realignment scheme designed accordingly; this is followed by an EIA, which may identify aspects that require modification to the design, and so on.

The requirement for EIA may arise from two European Union directives, which are embodied in UK law, or through a series of regulations applying to different types of project and consent routes. In England and Wales one or more of the following may also apply:

- where planning permission is required, the applicable regulation is usually The Town and Country Planning (Environmental Impact Assessment) (England and Wales) Regulations 1999 (SI no 293)

- for flood defence projects not subject to planning permission, the Environmental Impact Assessment (Land Drainage Improvement Works) Regulations 1999 (SI no 1783) will normally apply

- for port-related projects, the Harbour Works (Environmental Impact Assessment) Regulations 1999 (SI no 3445) will normally apply.

Scotland and Northern Ireland have their own corresponding EIA Regulations (The Environment Impact Assessment (Scotland) Regulations 1999 and The Planning (Environmental Impact Assessment) Regulations (Northern Ireland) 1999). In each case, the Regulations define the characteristics of projects for which EIA is, or may be, required, the information to be submitted and the procedures for preparing and submitting an environmental statement.

Detailed guidance on how to undertake effective EIAs is beyond the scope of this publication, but there is a wide range of reference materials available on the topic from, among others, Defra.

2.5.2   Existing environmental assets

Assets that should be considered in assessing suitability of a site for managed realignment include the presence and quality of:

- nature conservation interest, including individual habitats, species and designated sites

- cultural heritage interest, including scheduled ancient monuments, listed buildings and archaeology

- landscape and visual amenity

- agricultural land and other land use

- fisheries

- recreational use

- navigation

- water resources, including aquifers and fresh surface water

- water quality, including salinity, visual appearance and bathing waters

- the social importance of the site.

The relevance of environmental assets is recognised in Defra's current PAG notes for coastal defence schemes (www[13]). Sustainable conservation of SSSIs and other designated areas, scheduled ancient monuments and listed buildings attracts additional points within the priority scoring system, while sustainable conservation of SPAs, SACs and Ramsar sites is considered a legal requirement and should be approached *outside* the priority scoring system. Creation (net gain) of biodiversity action plan habitats (which includes saltmarsh, mudflat and vegetated shingle) also attracts additional points. Conversely, schemes that result in a net loss of habitat will score zero for this criterion.

Conservation of many environmental assets is also a material planning consideration under various national Planning Policy Guidance (PPG) notes and Technical Advice Notes (TAN, for Wales) (www[15]), particularly:

- PPG 7 *The countryside – agricultural land of Grades 1, 2 and 3a*
- PPG 9 *Nature conservation – SSSIs, SPAs, SACs and Ramsar sites*
- PPG 15 *The historic environment – listed buildings, conservation areas, historic parks and gardens and the wider historic landscape*
- PPG 16 *Archaeology and planning – scheduled monuments and non-scheduled archaeological sites*
- TAN 5 *Historic buildings and conservation areas 1996*
- TAN 6 (W) *Agricultural rural development 1996*
- TAN 6* *Consultation draft for archaeology 1996*
- Circular 61/91 *Planning and the historical environment: historic buildings and conservation areas*.

PPGs 15 and 16 are no longer in force in Wales, having been superseded by Planning Guidance Wales (which has supplementary planning advice). Other relevant policies include planning policy statement relating to AONBs and national parks and the Countryside Rights of Way Act and Wildlife and Countryside Act.

Protection of assets *in situ* should take place only where it is sustainable to do so. Government regulators accept that in many cases conservation *in situ* will not be sustainable and in such cases assets should be re-created elsewhere.

Many of these policies are incorporated into statutory structure and local plans. Where managed realignment would adversely affect such assets, this could represent a significant obstacle to gaining planning permission or may lead to a range of conditions being imposed.

Nature conservation

Avoiding realignment over sites of existing wildlife value may be hard to achieve because of other over-riding factors such as sea level rise. It may be necessary to make a strategic decision that freshwater or terrestrial habitat will need to be replaced elsewhere to make way for intertidal habitat in the interests of sustainable coastal or estuarine management (see Doody, 2001, for an ecological perspective). The statutory consultees on such matters are the country conservation agencies: English Nature (EN), Countryside Council for Wales (CCW), Scottish Natural Heritage (SNH), Environment and Heritage Services Northern Ireland (EHS). There may be opportunities to maintain reedbed and grazing marsh within a realignment area or to re-create lost habitat in other more sustainable locations, which is likely to include sites not immediately adjacent to the coast, for example in river floodplains. Brackish and saline lagoons may present a particular issue, as they are rare in terms of the total area of resource, identified as priority habitats by the Habitats Directive, and can only exist adjacent to the shoreline (Bamber *et al*, 2001). Lagoons are often separated from the sea by a barrier beach, sand dune or seawall and it may be possible to design such habitat into the new realignment scheme. Defence realignment may cause localised changes to intertidal habitats in the vicinity of a proposed breach. The extent of any effect will need to be assessed and can be limited by effective design. Depending on the resource and value of the intertidal, this may not be a constraint.

In some cases, features of geological or geomorphological interest may be compromised by, or benefit from, defence realignment. Adverse impacts are less likely to occur where retreat is to high ground, as landforms will have space to develop. Where the interest is in the relict features, recording and selecting a suitable route for any necessary new defence is likely to be the most effective approach to mitigation. It is also possible that the defence, in the form of shingle ridge or sand dune system, is also the feature of interest. In such cases allowing realignment to a more sustainable position (ideally by natural roll-back or, if not sustainable, by engineering a new position) may be acceptable on nature conservation grounds.

Landscape issues

A report for the Countryside Council for Wales (ECUS, 2003) discusses the importance of landscape issues in coastal defence design. Managed realignment can involve major changes in the appearance of the landscape. The advantages and disadvantages depend on which elements of the landscape will be lost, changed or created and the extent of these changes. Examples of some of the possible changes are listed below.

Potential advantages

- Replacement of hard engineering structures including coastal defences, and possibly roads, buildings, railways etc, with soft engineering shoreline layouts

- maintenance of some of the historic landscape such as old seawalls and drainage patterns

- alternative to construction of potentially unsightly hard defences in the long term

- restoration of original (natural) landscape providing a more dynamic environment

- widening of the foreshore and open space

- creation of habitat with increased landscape interest from changes in flora and fauna.

Potential disadvantages

- Change of existing landscape features such as defences, grazing marsh, golf links, sand dunes, timber groynes, cliffs etc

- construction of a secondary defence such as a flood embankment further inland, thus relocating the artificial landscape

- existing defences and other features might not be entirely removed

- construction of additional structures and features that are necessary to ensure that the scheme fulfils its technical requirements

- dead trees and hedges may be unsightly (and can also pose navigation hazards)

- large exposures of bare mud (mudflat or before vegetation colonisation) have often been considered unsightly, especially where the public are used to green fields

- realignment may change the location/extent of impact from wind-blown and/or water-borne litter.

It is therefore important to assess the impacts on landscape of a potential managed realignment site and to take into account the context of landscape issues in the scheme design by using appropriate structures. It is also important to consider from where people will view the realignment area. A site that can be seen only from ground level will have a different visual impact from one that is looked down upon from high ground. All AONBs have their own landscape assessments that can provide helpful guidance. Further guidance is available under the Countryside Character Initiative (www[1]) and there may be other local landscape assessments available for a particular area. The report for the Countryside Council for Wales (ECUS, 2003) states:

> ...as a general principle, defences that are artificial in appearance are more appropriate on developed coasts than on underdeveloped or natural coasts. Defences that simulate natural coast elements in appropriate locations generally have fewer adverse landscape and visual effects than defences that are artificial in their location and appearance.

Historic environment

There should be a presumption in favour of retaining scheduled ancient monuments and listed buildings *in situ*, unless this is shown to be unsustainable in terms of defence requirements. In exceptional cases physically relocating a cultural asset to conserve it outside a realignment area may be justified, although it is recognised that some of the interest relating to the original setting of the asset would be lost.

Non-scheduled archaeological sites should also be retained where possible, though where managed realignment would result in the change in a site then conservation by recording may be an acceptable alternative. In some cases, proposed managed realignment may affect a recognised historic landscape, within which existing and former seawalls provide evidence of the history of land enclosure, human occupation and past land uses. Managed realignment in this type of location would reverse that historic process and could lead to the loss of evidence. Research and documentation of such landscapes may be appropriate mitigation if combined with a watching brief during construction. Breach realignment may be appropriate where there is a need to retain some of the historic embankment sequence that was used in reclamation.

English Heritage (Trow and Murphy, 2003) states that it does not seek to place unnecessary constraints on schemes of economic and ecological importance. The archaeological process is iterative, involving stages of assessment, survey, evaluation, excavation/recording where necessary, and monitoring during and after construction. Where historic assets would be adversely affected by managed realignment, this could represent a significant obstacle to gaining planning permission unless evaluation and mitigation procedures are followed. Early consultation with local authority historic environment advisers or the Welsh Archaeological Trusts in Wales is advisable. Relevant legislation relating to the historic environment includes the Ancient Monuments and Archaeological Areas Act (1979), Town and Country Planning Act (1990) and the Protection of Wrecks Act (1973). The effects of scour and enhanced erosion, which may affect other archaeological sites, historic buildings or structures, may mean that post-scheme archaeological monitoring, and mitigation should be undertaken. Where historic properties, scheduled ancient monuments, and/or Grade i or ii listed buildings are likely to be affected by a scheme the relevant authority must be contacted (English Heritage, pers comm, 2004).

English Heritage outlines the following archaeological implications of managed realignment:

- where early timber and earth seawalls are the subject of realignment, sampling for dendrochronological, palaeoecological and wood technology analysis should be undertaken

- mudflat, lagoon or creek creation by lowering an existing land surface before defence realignment is potentially destructive of items of archaeological interests and should be accompanied by a watching brief, coupled with controlled excavation of any known sites. Ideally, historic creeks should not be excavated

- re-wetting of partly dewatered sediments may damage some archaeological materials and help preserve others; materials such as wood, metal and ceramics require further research to determine the effects.

If archaeological issues are likely to arise then it may be appropriate to appoint a specialist contractor to undertake a desk study and walkover survey. Depending on the findings, the survey may need to be followed by excavation or a watching brief during construction. Tidal exchange schemes may have fewer archaeological implications than other forms of managed realignment because they can have less direct disturbance to the existing defence (compared with breach or bank realignment) and inundation can be modified to control the area (depth) and frequency of flooding.

Box 2.2   *Example of historic/archaeological site and managed realignment*

> No archaeological sites were known in advance at this proposed managed realignment site, but a programme of archaeological recording was agreed between the county archaeologist and the Environment Agency. Following the planning application, and before construction, a borehole survey and assessment was commissioned and it showed a pre-transgression land surface and identified the areas most suitable for historic settlements. Archaeological recording was undertaken during construction and a five-year-long monitoring programme followed to detect and record any new interest exposed by on-going scour. This was subsequently extended by a further three years.

Agricultural land

Agricultural land is taken into account in determining the economic case for defence realignment against other options. Protecting Grade 1, 2 and 3a agricultural land from development is also a material planning consideration within national (PPG 7 *Countryside*, Paragraph 2.17) and many local planning policies. On the other hand, Defra offers the opportunity for stewardship payments for converting agricultural land into intertidal habitat. Such land use change will be considered under any planning application and needs to be informed by a strategic approach to estuaries/coasts.

Fisheries

New wetlands created by managed realignment can provide important fish spawning/nursery areas, thereby benefiting fisheries sustainability. They can also provide shellfishing opportunities. However, managed realignment may adversely affect fisheries, in particular existing shellfish beds, if there is a significant change in scour and/or sedimentation or if contaminants are present in the realignment area. Scour arises from the progressive enlargement (up to a point of equilibrium) of the tidal creeks conveying water to and from the managed realignment sites or the discharge across the foreshore and can result in shellfish beds being physically washed away. Contamination, if present, may be diffused into the environment and affect local fisheries interests. There is also potential for increased sedimentation at the fishery sites as a result of sediment being mobilised and redeposited, which may have implications under the Shellfish Waters Directive (see Section 1.2.4). These impacts may occur over a range of timescales, but rapid change is likely to cause the biggest problem. Even a short-term smothering of sediment can damage fisheries. It is therefore important to establish the presence and extent of all such fisheries within the vicinity of managed realignment, not just those immediately adjacent to the realignment, and to consider possible short-term and long-term changes.

Detailed assessment of the potential for scour and sedimentation is needed, including, where appropriate, hydrodynamic and geomorphological modelling. These techniques are unlikely to be able to predict all the effects, however, so it may be necessary to allow for monitoring and mitigation measures or financial compensation to owners if subsequent adverse effects on fisheries develop that are attributable to the realignment. Mitigation can be incorporated into realignment schemes; for example, at one site the identification of sedimentation as a potential issue, through consultation, led to the decision to alter the original design concept to leave a sill to trap sediment within the site and reduce tidal velocities outside the site.

Recreational use

Consideration needs to be given to the recreational issues associated with the site. The impacts on existing usage and the potential for enhancement or loss following the implementation of the scheme may be critical in achieving stakeholder acceptance. Managed realignment schemes offer scope for a wide range of recreational usages, and it is important to consider them all as well as to identify existing recreational activities and how these might change as a result of realignment. Problems have occurred, for example, where footpaths are to be extinguished and re-routed to a less favourable location with less attractive views.

Navigation

Managed realignment may also have an effect on navigation by causing scour or deposition in the deep-water channels and altering flow patterns. Scour may not be a problem unless it results in a change of alignment of the channels, whereas deposition could lead to the need to dredge the channels or to increase the frequency of dredging.

Detailed study of the potential for affecting navigable channels should be undertaken before implementing managed realignment including, where appropriate, numerical or physical modelling. These impacts might represent a constraint to implementing realignment or require appropriate continuing management.

Water resources and quality

Freshwater abstractions, whether from surface or groundwater, are a valuable resource. If they are lost or compromised as a result of a deliberate act of defence realignment, there may be an obligation to financially compensate the owner of the abstraction rights. The most obvious effect of realignment is likely to be salinisation of surface and groundwater farther inland and should be identifiable on a groundwater protection zone map. It is recommended to discuss groundwater issues with the Environment Agency or SEPA, but, where saline intrusion is identified as a problem, installation of an impermeable cut-off will help protect the resources. The extent of any effects will depend on the existence of underground structures, the geology and particularly the permeability of the underlying strata.

Impacts on water quality must also be considered, for example bathing water quality.

Target habitats

Depending on the scheme objectives, managed realignment may (but does not have to) aim to create a specific target habitat. The target habitats of many realignment schemes are often saltmarsh and intertidal flat but can also be lagoons, shingle ridges, sand dunes and transition habitats (see Appendix 2). Emphasis is given to saltmarsh and intertidal flat areas because:

- saltmarsh and intertidal flat vegetation and associated topographic features (creeks, salt pans, mud mounds, ridges and runnels etc) attenuate wave energy under low to medium tidal inundation (Cooper, in press) and therefore can reduce routine maintenance expenditure on coastal defences. The attenuating effect of a saltmarsh is dependent on the maturity of the marsh surface and its vegetation cover. Newly created marshes or seasonally vegetated mudflats will not be as effective as mature marsh systems (Möller *et al*, 1996, 1999, 2001, 2002)

- saltmarsh is valuable to biodiversity, both in its own right and as a feeding and roosting area for wildfowl and wading birds, for its biological productivity and fisheries and it is a priority habitat in the national and many local biodiversity action plans

- saltmarsh is often the type of landscape most affected by coastal squeeze and is therefore declining as a result of the presence of existing defences

- saltmarsh tends to accrete intertidal sediment and, to some degree, can therefore respond to sea level rise (French and Burningham, 2003)

- many areas over which realignment is most likely to take place have, historically, been saltmarsh.

Different saltmarsh plants have different tolerances to inundation frequency (and associated other factors) so that well-developed saltmarshes exhibit several zones at different elevations. Elevation is not the only controlling factor on whether vegetation can establish and support itself, but the level below approximately 450 tides a year will tend to form mudflat, where eelgrass (*Zostea spp*) or algal communities might occur (eelgrass is rare on many coasts around the UK). The level between about 450 tides and 360 tides a year can support "pioneer" saltmarsh species such as annual samphire (*Salicornia*) and sea aster (*Aster*).

Figure 2.3   *Pioneer species inhabit the lower zone of natural saltmarshes and will be early colonisers (probably the first or second year after realignment)*

The pioneer zone leads into the lower and mid saltmarsh where inundation might typically be less than about 300 tides a year and species such as common saltmarsh grass (*Puccinellia*) and sea purslane (*Halimione*) can be supported. Around mean high water level, the mid saltmarsh, which includes species such as sea lavender (*Limonium*) and red thrift (*Festuca*), grades into the upper saltmarsh, which can support species such as sea couchgrass (*Elytrigia*) and shrubby seablight (*Suaeda Vera*). Driftline communities can occur around the extreme high water spring tide level and grade into the transitional habitats that occur up to above the highest astronomical tide (HAT). The upper limit of saltmarsh is technically defined as 1 m above HAT level, but this may underestimate the extent. An upper limit might also be defined by multiplying HAT by 1.23, above which terrestrial vegetation occurs (Burd, 1995; Hough *et al*, 1999). The actual species and extent of a saltmarsh vary depending on the actual local tidal range and factors such as sediment and nutrient supply and water clarity (for photosynthesis). In Poole Harbour, for example, which has a very small tidal range, saltmarshes occupy a slightly lower position in the tidal frame than in other UK estuaries (Rodwell, 2000).

Many saltmarshes do not have the full range of zonation or lead into terrestrial vegetation, typically because they are backed by a road, field or sea defence, which limits the opportunities for saltmarsh succession. The upper part of saltmarsh should naturally integrate to terrestrial vegetation, but this transition zone is rare in the UK.

Figure 2.4   *The full range of saltmarsh zonation is rare, but managed realignment offers the opportunity to provide this (for this site, zonation developed within six years of realignment)*

At low elevation, in relation to the tides, mudflats or sandflats will develop with an intrinsic value. They can be extensive and contribute substantially, in some areas, to the wilderness quality of the coastal landscape. They can support significant biomass resources (including invertebrate fauna) and these in turn provide food for large populations of waterfowl, wildfowl and waders. The high productivity associated with the tidal waters and flats also helps to support a number of spawning and juvenile stages of some commercial fish species. Intertidal flats are often considered easy to create, but successful habitat depends on the presence of a suitable substrate type and sediment supply. Sand and soft mud will support quite different faunal communities, while highly compacted mud is likely to support fewer organisms in the first instance. This may change over time. For example, inspection of London Clay in a landward borrow pit (with a leaking tidal flap) demonstrated that ragworm and lugworm will

PART II

burrow into inundated clay substrate and so begin bioturbation. Where less compacted sediment is available, however, these areas will be used preferentially.

Maintaining or extending the width of intertidal flat can also help attenuate wave and tide levels and help in the functioning of beach and ridge formations. It may be possible to make some predictions of the distance inland required to accommodate sea level rise, by using techniques such as the Bruun Rule (see Appendix 2). The reduction in waves that might be afforded by a wider intertidal profile can help protect the landward part of the intertidal profile and this can allow vegetation to be maintained or established. The intertidal zone (vegetated at high elevations, unvegetated at lower elevations) can be viewed as one morphological element with close linkages and sedimentary feedback mechanisms across it.

Other coastal habitats that may benefit from managed realignment include vegetated dune ridges and shingle ridges. Sand dunes are formed of grains generally between 0.125 mm and 0.25 mm in diameter (particles smaller than 0.1 mm and greater than 1.6 mm are relatively uncommon as wind-blown material) and moderately well sorted, and shingle ridges are formed of sediment generally greater than 2 mm in size. Little guidance can be offered in respect of this type of environment except to apply expert geomorphological assessment and consider the methods by which the different sediments are transported. Sand dunes are reliant on a supply of sand from the foreshore and hence require wind as the principal driving force and the sediment drying sufficiently to be transported. Shingle ridges will require a supply of shingle – which in many cases may be reworked relict sediments – and wave action to drive up a ridge (Doody and Randall, 2003). Where there is mixed sediment, or where sand and shingle are juxtaposed, it can be very difficult to assess what habitat may form or survive. A mixed sediment environment is likely to change and respond to storm events such that habitat changes will also occur. Shingle ridges may be formed under storm conditions and subsequently reworked and redistributed. Shingle ridges may also form under a storm and then form the nucleus of a sand dune, which is destroyed in a following storm. Such dynamic change is natural but difficult to predict.

Dunes and shingle ridges are more likely to be the subject of habitat restoration than creation of new areas. Instances where managed realignment may be applicable include where a dune or shingle frontage is presently being managed. This may take the form of groynes, a seawall, beach recharge or beach profiling, to limit erosion or reduce flood risks to the hinterland (Simm, 1996). The result of such interventions is often that the beach habitat is degraded, functional links between beach and dunes or ridges are disrupted, and the features of landscape or conservation interest, such as vegetated shingle or embryonic dunes with pioneer vegetation, are either missing or not in favourable condition. Realignment may entail constructing a new line of defence to the landward and allowing the dune or shingle frontage to respond to natural processes, including sea level rise. This may lead to an improvement in its conservation and flood defence value.

2.5.4 Infrastructure and landownership

The minimal presence of property and built infrastructure in an area is desirable if realignment is being considered. Types of assets that need to be considered include residential and commercial property, roads, railway lines, power cables and pylons and telephone lines (particularly where supported by structures within the proposed realignment area and/or underground cables). It may be possible to overcome constraints by relocating the infrastructure or providing a counter wall or localised protection. This will need to be taken into account in the scheme economics (see Section 2.6).

Land ownership is a significant factor in determining the feasibility of successful managed realignment. Negotiations are likely to be easier where there is one or only a few landowners within a proposed site. Their support, or otherwise, for a realignment scheme will also be a significant factor in success or failure. This is likely to depend on how the proposal links to a strategic approach, how the proposal is presented, whether it includes an offer to purchase land within the realignment area or some other form of compensation, and how the community perceives it. Managed realignment is often highly politically sensitive, so carefully planned and thorough community involvement should form a part of the project (see Section 1.5).

2.6 ECONOMIC VIABILITY

The economic viability of a managed realignment scheme must be considered in terms of the whole-life costs (capital and maintenance costs, usually reduced to present values). Adopting a holistic approach to economic appraisal allows the long-term maintenance and capital costs to be identified and these can be reduced by adoption of the most sustainable strategy.

In order to be eligible for flood management funding from Defra, an economic appraisal of the scheme costs and benefits should be completed in accordance with Defra's Flood and Coastal Defence Project Appraisal Guidance (Defra 2000, www[13]). Managed realignment will be considered economically viable only if the benefits outweigh the costs or if it provides the least-cost means of meeting a legal requirement. The benefit/cost ratio should be used in the decision-making process to compare the economic viability of managed realignment options with alternative environmentally acceptable and technically sound options. Environmental costs and benefits should be described and considered either within or alongside the economic appraisal. Defra's PAG3 (economic appraisal) and PAG5 (environmental appraisal) provide guidance on how this should be done.

In addition to project appraisal, Defra applies a system of prioritisation to help decide which projects are eligible for grant aid. This is necessary because the demand for investment usually exceeds available resources. Defra's prioritisation system takes account of social and environmental benefits as well as economic efficiency. Schemes that are necessary to meet the requirements of the Birds or Habitats Directives are fundable by Defra grant aid outside of the prioritisation system.

2.7 CONSENTS, PERMISSIONS AND LICENCES

As with any form of change to the coastline, a range of consents, permissions and licences is needed to proceed (Halcrow Consulting Engineers, 1994, revised 2001). The approach to consents and licences needs to be programmed into a managed realignment scheme from the outset. To apply for these consents and licences the necessary studies, investigations and analysis will need to have been undertaken to appropriate standards and/or levels of robustness. The sequencing of activities should be considered, because information from one activity often links to another. For example, the outputs from hydrodynamic and geomorphological assessments will inform the outline design and assessments of ecological and landscape change; all three will feed into redesign (if necessary) and form a part of an environmental impact assessment (see Section 2.5.1).

Survey work needed to inform a licence submission may have to be undertaken at particular times of the year. Ecological survey, for example, usually takes place between March and September, while an over-wintering bird survey is generally carried out between October and March with core counts occurring between November and February. If direct survey is not needed, because there suitable data and information is already available, then the process may be streamlined. It is important, therefore, to set out all aspects of the consents and licences as part of the project planning and to identify critical information linkages. The time for consultations, possible revision or redesign and resubmission should be included in the programme, together with the costs of applications, which will be incurred whether they are successful or not. The particular aspects that might cause iteration in the process cannot always be identified before investigations take place and in some cases additional survey investigation or analysis has been required as well as redesign of the scheme. The iterative nature of the process will make it necessary to return to those who have already been consulted with any changes proposed to the scheme.

The application process can be time-consuming. Project experience to date suggests that, once the data and information have been gathered, it can take between nine and 15 months to secure all the necessary consents and licences. Dependent on the iterations in the process, this time-scale may be extended. A linear flow through the process is uncommon, as those reviewing the information are often unfamiliar with managed realignment and adopt a precautionary or cautious approach, seeking clarification on issues and requiring resubmission of information. This period can be reduced through effective consultation and communication with those who will be giving the consent, permission or licence during the initial investigations and studies. It is also anticipated that this time will reduce where the consenting authorities have dealt with previous managed realignment schemes.

The following sections detail the types of consents that may be required for a managed realignment scheme. The responsible authorities need to be approached to determine the need for a consent or licence and to identify specific information needs. See also Section 2.5.1 on environmental impact assessments; Although an environmental impact assessment (EIA) is not a consent as such, it plays an important role in the issuing of a number of consents including planning permission under the Town and Country Planning Act.

2.7.1   Planning permission under the Town and Country Planning Act 1990

In general, development landward of the mean low water mark, including coastal defence schemes, requires specific planning permission from the local planning authority. There is an exception, in that improvements to existing flood defence schemes carried out by the Environment Agency normally receive deemed planning permission under the Town and Country Planning (General Permitted Development) Order 1995, which means that specific planning permission is not required in such cases. However, this exemption rarely applies to managed realignment schemes that entail constructing a set-back defence on a new alignment and/or a change of land use in the retreated area. Where one of these features applies, or there are other factors such as European designations that mean permitted development is not applicable, then specific planning permission will be required. Planning consent will also be needed if a body other than the Environment Agency intends to remove an existing flood defence. Furthermore, the Habitats Regulations (1994) introduce a requirement for planning consent for schemes that would otherwise be permitted development but which may adversely affect the integrity of a European site. Where the integrity will be adversely affected, consents are subject to approval by the Secretary of State, which can incur delays of the order of a year or more, in addition to the normal planning process. In Scotland planning consent under the Town & Country Planning (Scotland) Act 1997 is required for defence construction schemes above MLWS.

2.7.2   Habitats Regulations

Under the Conservation (Natural Habitats & c) Regulations 1994 (SI no 2716), known as "The Habitats Regulations", projects that may significantly affect a European site (SAC or SPA), whether or not they are actually located within the site boundary, require an "appropriate assessment". This provision also extends to candidate SACs in England (but not in Wales, Scotland or Northern Ireland) and Ramsar sites. The country conservation bodies can advise whether this applies in a particular case and should be consulted at an early stage. The appropriate assessment itself, however, is undertaken by a "competent authority", which the Regulations define as a body from which consent is required. Where the development is not otherwise subject to consent, the competent authority is the local planning authority. If the appropriate assessment finds that the proposal may adversely affect the integrity of the European site, it can only go ahead through a full planning application and if there no alternatives and imperative reasons of over-riding public interest exist. For European sites that host a priority habitat or species, imperative reasons of over-riding public interest can relate only to human health, public safety, or beneficial consequences of primary importance to the environment. Where a European site is not host to a priority habitat or species imperative reasons of over-riding public interest may be of a social or economic nature. The Secretary of State/National Assembly of Wales is responsible for determining applications for imperative reasons of over-riding public interest and, if such reasons are proven, must ensure compensatory measures are secured.

2.7.3   Wildlife and Countryside Act 1981 (as amended by the Countryside and Rights of Way Act (CRoW) 2000)

The Wildlife and Countryside Act 1981, as amended by the Countryside and Rights of Way (CRoW) Act 2000, provides that consent must be sought from the country conservation bodies where it is proposed to undertake a permitted development that would affect a site of special scientific interest. The country conservation bodies can advise whether this applies in any particular case and should be consulted at an early stage. The CRoW Act also introduces a requirement for positive management by SSSI owners and occupiers (incorporated into non-technical "Views about management" prepared by the country conservation agencies) and a duty on public authorities to enhance conservation within SSSIs in the exercise of their functions. It also gives a duty for public bodies towards AONBs, under Part IV of the CRoW Act, when undertaking any project that may affect part of an AONB.

2.7.4   Coast Protection Act (1949)

The Coast Protection Act 1949 (Coast Protection (Notices) (Scotland) Amendment Regulations 1996 and Coast Protection (Notices) (Wales) Regulations 2003) accompanied in Scotland by the Flood Prevention Scotland Act 1961 (as amended by the Flood Prevention and Land Drainage (Scotland) Act 1997) allow local authorities to promote appropriate schemes on land for coast protection or flood prevention works where this is deemed necessary for the wider public interest. Authorities are also permitted to undertake maintenance and emergency works. If a coast protection scheme receives any public funding under the Coast Protection Act it will need consent from Defra Flood Management Division under Section 5 of the Act.

Where coastal defence works are carried out below the level of mean high water spring tides, consent will be required from the Department of Transport under Section 34 of the Coast Protection Act 1949 (as amended by Section 36 of the Merchant Shipping Act 1988). This applies to the construction, alteration, improvement, removal or deposit of

PART II

any objects or materials anywhere in UK territorial waters. The main purpose of the Act is to restrict works that may be detrimental to the safety of navigation, but additional environmental restrictions must also be taken into account. Consents under the Coast Protection Act are administered by the Marine Consents and Environment Unit (www[6]).

2.7.5 Food and Environment Protection Act (1985)

The Department for Environment, Food and Rural Affairs (Defra) in England and the National Assembly in Wales have a statutory duty to control the deposit of articles or materials in the sea and tidal waters, the primary objectives being to protect the marine ecosystem and human health and minimise interference and nuisance to others. Environmental considerations include the potential hydrological effects, interference with other marine activities, the possibility of turbidity and drift of fine materials to smother seabed flora and fauna, and adverse implications for designated conservation areas. Licences under Part II of the Food and Environmental Protection Act 1985 are administered by the Marine Consents and Environment Unit (www[6]).

2.7.6 Environment Agency and Scottish Environment Protection Agency

The Environment Agency administers consents under the Environment Act (1995), Land Drainage Act (1991), and Water Resources Act (1991). This includes responsibilities for flood defence, water resources and water quality. It may also be the responsible authority for overseeing the implementation of Flood Defence (Land Drainage) Bylaws/Sea Defence Bylaws and ensuring application of parts of the Highways Act (1980, overseen by local authorities). The Environment Agency is also an adviser on flooding and flood risk issues and the lead government body for saltmarsh and mudflat under the UK BAP. The Environment Agency has a role in promoting sustainable forms of development in exercising its duties.

The Scottish Environment Protection Agency (SEPA) has some similar functions to the Environment Agency including responsibilities under the Flood Prevention & Land Drainage (Scotland) Act 1997 and the Roads (Scotland) Act 1984.

In general, flood defence works that may affect watercourses and/or existing defences require a single consent under these Acts (note this is not the only consent required) and applications should be made to the relevant Environment Agency or SEPA region (www[7]). It should be noted that any new line of defence needs to meet normal standards and be acceptable to the Environment Agency or SEPA.

2.7.7 The Crown Estate

The Crown Estate owns around 55 per cent of the foreshore (between mean high and mean low water) and approximately half of the beds of estuarine areas and tidal rivers in the United Kingdom. It also owns the seabed out to the 12-mile territorial limit and there are substantial areas of Crown Estate-owned land. Where permanent works or operations are proposed on the seabed, the Crown Estate needs to be consulted and where appropriate a licence obtained (www[8]). To date the Crown has not claimed ownership of new intertidal areas created through managed realignment but written confirmation of this should be sought. In some estuaries, agents, such as port and harbour authorities, act on behalf of the Crown Estate.

2.7.8   Port authorities

Realignment schemes within estuaries frequently require consent from the relevant port authority for direct works or for any potential impact to navigation. For example, a licence is required from the Port of London Authority before works are undertaken in, under or over the Thames seaward of mean high water and landward of the seaward limit (Foulness Point in Essex to Warden Point in Kent), or before the banks of the Thames are cut in any way (www[9]).

2.8   METHODS OF MANAGED REALIGNMENT

Managed realignment is usually implemented by one of two principal methods, namely breach realignment or bank realignment. Other techniques include the lowering of a section of existing tidal defence to produce a spillway or the installation of sluices or pipes in an existing defence to allow the periodic exchange of the tide (tidal exchange systems). A further scenario is "do nothing", although this would result in unmanaged, rather than managed, realignment (see Box 2.3).

Box 2.3   *Limited intervention/do nothing – unmanaged realignment*

Under this strategic option no capital or maintenance works are carried out either on the existing defence or within the hinterland behind the defence, except to meet health and safety requirements. At some time in the future the tidal defence will fail in certain locations; the timescale being dependent upon the residual life of the existing defence and the forcing parameters to which it is exposed. This could be described as "unmanaged realignment" Initially, failures are likely to create weirs rather than full breaches. The development of breaches may concentrate flows, causing negative impacts (scour and hydraulic force) on other defences, navigation and the environment.

The subsequent development of the site is likely to differ from that under a managed realignment option, as breaches would occur in undetermined locations and to unpredictable heights and widths. There is therefore likely to be considerable uncertainty in the outcomes and the impacts on property and the environment inside and beyond the site.

If rising ground exists behind the tidal defence, and the extent of the inundated area is limited, this may be an acceptable risk. Indeed, early work on managed realignment assessed how "do nothing" sites had responded following seawall/embankment failure (Burd, 1995).

2.8.1   Breach managed realignment

Breach managed realignment comprises the removal of certain lengths within an existing defence. A new defence alignment may be created some distance landward of the existing defence by the construction of a new structure or by naturally occurring high ground. A combination of the two may also be used.

The technique is well suited to schemes where the primary purpose of the works is the rapid creation of a definable new intertidal habitat for environmental benefit. It is appropriate where a measure of wave protection is needed for the site and/or realigned defence and may reduce the extent of any scour protection required. Over time, development of vegetation and accretion of sediments within the new intertidal area may reduce the need for the physical protection of the realigned defence.

Breach realignment is suitable where sensitive local issues or potentially adverse effects of scheme implementation have been identified, for example erosion of downstream tidal defences within an estuary or detrimental effects on faunal communities due to increased turbidity. Carefully evaluating the number of breaches and their width(s), and implementing an appropriate monitoring programme, can manage these effects.

Cost comparison

Breach managed realignment may be more cost-effective than bank managed realignment, because part of the existing tidal defence is left and allowed to deteriorate naturally over time, rather than having to be physically removed. The debris from such deterioration may cause safety problems, however, as well as being visually obtrusive, particularly if the embankment is armoured.

Figure 2.5   *Breach managed realignment can be cost-effective, removing only section(s) of the existing defence*

2.8.2   Bank managed realignment

Bank managed realignment can comprise the removal of an entire length of existing tidal defence such as a seawall, a flood embankment or a shingle ridge. A new alignment may be created some distance landward of the existing defence by the building of a new structure or by using existing high ground. Different implementation techniques can be employed for different purposes, but it is commonly recommended in the "general" literature that complete bank removal is the form of technique that least interferes with natural processes. It is more expensive than breach realignment, which may preclude its adoption for some schemes. The feasibility of safely removing extensive lengths of banks also needs careful consideration, as once tidal inundation occurs site working can be considered to pose an unacceptably high level of risk.

Shingle ridges or sand dunes

Bank managed realignment can also comprise the
moving inland of soft defences such as shingle ridges or sand dunes. These types of defence (and habitat) should be rolled back inland under natural processes where possible. Most currently managed ridges do migrate inland over time, but the rate of movement has been arrested. It may be necessary to modify the profile of a managed ridge to achieve sustainable roll-back. Where the ridge is in an unsustainable position it may be necessary to move the ridge inland to allow this process to take place effectively. This may require de-stabilising the existing ridge by physical means or removal of sand dune vegetation. As with the removal of hard defences, it may be possible to allow the system to function with inundation to high ground or it may be necessary to provide a new, inland, defence line.

Where management practices or changes to natural processes (such as reduction in sediment supply alongshore) have created an unsustainable position for sand or shingle ridges, relocation inland may prove more sustainable. Allowing the relocation of, or physically relocating, a ridge inland may reduce the wave energy to which it is exposed, or, in the case of sand dune ridges, provide a wider foreshore for wind-blown sediment. A new alignment may be necessary to reduce the risks of catastrophic breaching and provide the opportunity for a more self-sustaining system.

Intertidal areas

Bank managed realignment creates a wide intertidal profile, which serves to dissipate wave energy in an effective and responsive manner. The technique is well suited to schemes where the primary reason for implementing the works is to
provide a better overall tidal defence system through the provision of wide, dissipative intertidal surfaces. The desire to maintain a beach may also be a driving factor under rising sea level.

A new intertidal area may be subject to regular wave action and this may be beneficial for ridge development, although it may also limit the rate of sedimentation of fine sediments. The technique can be suitable for the creation of habitats that exist at lower levels, such as mudflat or pioneer saltmarsh, where wave action is reduced. It can also provide a wider intertidal sandflat to feed sand dunes or beaches or provide protection to saltmarshes higher up the intertidal profile.

Cost comparison

Compared with breach realignment, bank realignment is more costly. In the case of natural ridges, some modification to the whole ridge might be necessary to allow roll-back and avoid seaward drawdown of sediment increasing the flood risk. Where it is a hard defence, it may be an important method to avoid concentration of flows exiting a realignment site, or to allow the wider functioning of an estuary system. On the open coast it may provide the opportunity to develop a stable embayment
morphology that is better suited to sea level rise conditions.

2.8.3   Tidal exchange systems

Normally, neither breach nor bank managed realignment include any further control of the tidal inundation of the site. Tidal exchange systems allow the volume of tidal flooding and number of tidal inundations of a site to be artificially controlled. The techniques may be applied as an interim measure on sites that are determined to be too

PART II

low in the tidal frame if exposed immediately to a natural flooding regime. The process of building up sediment in the site and allowing this to de-water and stabilise ("warping", used originally as an agricultural practice) can provide pre-treatment for breach or bank realignment. This may take some years and require continuing management, however. This method is also suitable where the existing wall needs to remain as an access route or because of its inherent heritage interest. Flood defence costs may be substantially reduced, as a reduced standard of flood protection will probably be required for the newly inundated land area.

Tidal exchange systems provide an option for private landowners to create small areas of habitat that can easily revert to other uses in the future. This technique can allow a defined depth of inundation on a site to create particular habitats for target bird or plant species, or potentially to avoid inundation of archaeological material. It has even been suggested that tidal exchange offers the possibility of tidal power generation as part of a managed realignment scheme.

Figure 2.6   *Tidal exchange systems have been used by private landowners to create small areas of habitat landward of the flood defences, leaving the defence in place*

The control systems may involve lowering part of the crest of the existing tidal defence to create a spillway or the installation of culverts or pipes through the defence through which the flow is regulated by penstocks or sluice (flap) valves. Where tidal inundation of a site is by a spillway, a means of draining the site at low water may be required unless a tidal lagoon is desired (Bamber *et al*, 2001). The relatively small hydraulic capacity of spillways, culverts and pipes compared with defence removal or breach creation has tended to restrict their use in the UK to managed realignment sites of only a few hectares, although in the Netherlands and Germany they have been used to control tidal flooding and encourage sedimentation over extensive areas. A further constraint is that the existing defence line has to be maintained (even at a lower standard) for as long as the spillway, culvert or pipe is to function, so potential defence cost savings associated with breach or bank retreat may not be realised.

Guidance on the use of tidal exchange for the creation of saltmarsh and mudflats is included in a review undertaken for the Royal Society for Protection of Birds and the Environment Agency (Lamberth and Haycock, 2002).

TIMING AND TIMING CONSTRAINTS

As most managed realignments are likely to involve earthworks, some of them substantial, construction would normally be undertaken between April and October when ground conditions are more favourable and less wave action is likely. Outside of this period accommodation works such as service diversions could be carried out. New defences may need to be built incrementally to avoid heave and ground-loading problems and to allow stabilisation of the placed sediment before the site is inundated. Depending on the amount of loading or the nature of the material to be placed, this might need to be undertaken over more than one season.

Consideration also needs to be given to the tidal conditions (in relation to the existing land levels) for the site. In some instances it will be possible to remove walls on neap tides without the land being inundated. This makes working conditions safer and can provide some days as a working window. There is also an advantage in breaching defences on neap tides as the water levels will increase gradually towards the spring tides, allowing any wildlife in the site to escape and the tidal prism to also increase gradually. This allows some period of adjustment in the system and the opportunity to monitor immediate impacts (and, if necessary, take remedial action). If the site is low in relation to the tide then breaching may be required on a single tide. It may be possible to reduce overburden and make all preparation in advance of the breach to minimise the works on the day of breach. In this circumstance it may be necessary to breach on a spring tide because the tidal range is increased and may allow works to be carried out over a longer period, following the falling tide. This will need to be assessed on a site-by-site basis and should be discussed and agreed in the method statement with the earthworks contractor. Care should be taken to assess daily conditions, as tidal surges may cause problems where partial works have been completed. A short-range weather forecast and tidal prediction should be undertaken to minimise this risk (www[14]).

PART II

Figure 2.7   *Reduction of flood defence in stages. It may be possible to reduce a flood defence in stages, having assessed tide levels and the risks to construction activities*

Depending upon the environmental designations, winter working in an estuary would not usually be permitted so as to avoid disturbing over-wintering/feeding birds. However, it may often be possible to construct any new realigned tidal defence without environmental time restrictions because the defence would usually be outside the area of the environmental designations and partially shielded from the estuary itself by the existing tidal defence. It has been usual to construct the realigned defence in the first year and breach or remove the existing defence in the second year. This allows the realigned defence to become fully established (and vegetated) before it is exposed to tidal inundation.

Experience and site monitoring has shown that on average, following breaching or removal of defences, it takes about:

- two years for soil chemical stabilisation and plant colonisation to occur
- four years for the significant establishment of annual plants and invertebrates
- up to five years for perennial plants to establish.

It is postulated that it will take around 10 years for a fully functional habitat to develop, and probably even longer for the hydrodynamics to reach some form of equilibrium.

A scheme for a breach in a seawall can take between three and five years to implement, patience and managing stakeholder expectations are therefore important (see Section 1.5). The needs of the scheme may be revisited to refresh the objectives and bring the focus back to achieving the intended benefits. It must be remembered that, for the environment, intertidal habitat cannot be made inland, so the coastal/estuarine location is critical, and that sites might take time to develop.

3 Information to support evaluation

Evaluation of a managed realignment option is important to determine whether it will meet the criteria for the aim and purpose of the project. The appropriateness of a scheme can be evaluated using this chapter. Proportionality plays a large part in deciding the approach to evaluation and what scheme might be suitable. It is also advisable to use other R&D projects (such as those described in Section 3.2) to assess and select the preferred managed realignment method.

3.1 PROPORTIONALITY

Information to support the planning and design approach needs to be collated to provide an appropriate level of information proportional to the project. It is important to identify the scale and level of detail in the work to be undertaken in proportion to the size of the scheme and its potential consequences.

Schemes of a few hectares, with little change in level compared with the seaward environment, will still require consents and licences but may be accepted as low risk using existing information only and professional judgement. The driver for the scheme is also relevant to the amount of study required. If managed realignment is to compensate for a particular loss under the Habitats Regulations, there will be a need to prove it can provide a suitable replacement. Such managed realignment would need more detailed work regardless of the site area, compared with a farmer applying for change in use under a countryside stewardship approach. To ensure proportionality, discussions must be held with the regulatory organisations responsible for the relevant consents and licences as part of a scoping process. The responsible authorities take a view on the requirements of a scheme and where the impacts are likely to be. These organisations should be able to advise what should be addressed and where, depending on the size of the site and the drivers involved.

3.2 NATIONAL R&D AND OTHER SUPPORTING STUDIES

It is important to refer to national R&D and other supporting studies to ensure that the appropriate method is used. The following are identified as key studies that should be noted when looking to carry out managed realignment.

3.2.1 Habitat R&D

FD1917 – Suitability criteria for habitat creation

The current Defra/Environment Agency study FD1917, "Suitability criteria for habitat creation", is intended to further the understanding of habitat creation from coastal realignment and provide guidance to engineers and managers on the selection of suitable sites within a given estuary. The specific purposes of the research are:

- review of existing knowledge and understanding of the criteria that influence the growth of natural saltmarsh and intertidal habitats
- review of the selection procedure for sites for habitat creation
- identification of parameters that can describe potential realignment sites with regard to habitat creation, including dependencies

- production of an electronic decision tool (influence diagram) and whole estuary GIS screening tools for users to assess the potential of specific sites for habitat creation schemes.

The electronic tools aim to help structure the requirements, and features, that make areas more or less likely to establish and support a desired type of habitat. The tools will aid decision-making and the investigation of different Defra and EA strategies for integrating and meeting future flood and coastal defence commitments alongside habitat creation schemes through approaches such as managed realignment. The research is also likely to highlight future R&D needs in this field and provide a directed focus on their use in the decision making process.

FD1918 – Habitat quality measures and monitoring protocols

This contract started in February 2003, will conclude in 2004, and is being funded by the Environment Agency and Defra. The project is concerned with providing guidance for the monitoring of managed realignment and habitat creation sites. Such sites cover the intertidal regions of both estuaries and coastal zones and include saltmarsh and mudflat habitats. The overall aim of the project is to develop measures of habitat quality and monitoring protocols to implement these. The project will therefore provide guidance for:

- the collection of better data in terms of relevance, consistency and statistical validity (including both baseline and ongoing measurements)
- the assessment of the success of habitat creation schemes
- validating the effectiveness of mitigation schemes and assessing the residual impacts.

In addition, the project will provide guidance to managers to enable corrective action to be undertaken where habitat quality objectives may not be achieved, or to develop alternative quality objectives that better reflect the capacity or capability of the site.

Another study that covers coastal habitats is the EU LIFE Natura-funded "Coastal habitat restoration: towards good practice" (English Nature *et al*, 2003), which is an integral part of the "Living with the Sea" project.

3.2.2 Futurecoast

This Defra-sponsored research and development study (Halcrow Group, 2002a) collated and presented a wealth of information intended to provide a baseline understanding of physical processes and geomorphology around England and Wales. Particularly useful maps and reports covered the following topics:

- coastal geology
- seabed sediments
- seabed features
- hydrodynamics (tidal residuals, offshore wave data)
- conceptual understanding of the generic behaviour of individual coastal landforms, including a detailed assessment of sea cliff behaviour
- conceptual understanding of the behaviour of coastal systems at a range of spatial and temporal scales
- historic changes in shoreline position
- estuarine parameters such as tidal asymmetry and mouth width, based upon the JNCC Inventory of UK Estuaries.

The limitations of the Futurecoast study, however, are its focus on open coast systems (although it includes some useful estuarine information) and its geographical coverage of England and Wales only. Furthermore, although it presented material on coastal tendencies (eg erosion), rates of change were not quantified. Defra has issued copies of the Futurecoast project output on CD to all local authorities and Environment Agency offices in England and Wales.

3.2.3   Estuary Research Programme Phase 1

The UK Estuaries Research Programme Phase 1 (ERP1) began in 1997 and was completed in 2000 (EMPHASYS Consortium, 2000a). The project was designed to improve understanding of physical processes operating in estuaries and, in particular, of available modelling techniques that can be applied to assess estuarine processes and morphology. The final report from Phase 1 covers the availability, application and limitations of predictive tools. A key point arising from ERP1 is that the uncertainty in the various techniques and the complexity of estuary systems means it is often necessary to synthesise the results of various techniques so as to produce a conceptual model for understanding the geomorphological system (Townend, 2002). Building from this understanding it becomes possible to explore "what-if?" scenarios for development and change. The modelling techniques investigated in ERP1 included top-down models, bottom-up models and hybrid models. Some of these techniques can be applied to coastal areas.

In April 2002, the ERP Phase 1 Uptake project began, and was again an Environment Agency/Defra-funded project. The project's aim was to encourage wider uptake of the ERP project ideas and experience through a series of workshops and the establishment of a set of best practice guidelines. One of the deliverables from the EMPHASYS project was a database CD covering the main relevant physical characteristics in estuaries, such as bathymetry and flows, together with environmental and chemical characteristics such as ecology and nutrient concentrations. This database has been updated with new data. The new version of the database has been released into the public domain as the Estuaries Database 2003 (ABP Marine Environmental Research, 2003a).

ERP2 is an umbrella of projects running over a three to five-year research and development programme, beginning in 2003, totalling some £2.7 million. The core projects include:

- FD1905, Estuary process research project, completion due 2004

- FD2107, Development of hybrid estuary morphological models, completion due 2006

- FD2116, ERP2, Interpretation and formalisation of geomorphological concepts and approaches, completion due 2005

- FD2117, ERP2, Development and demonstration of system-based estuary simulators, completion due early 2006

- FD1911, Freiston shore-managed realignment, an Environment Agency monitoring programme, completion due 2007.

Other areas identified as potential projects include:

- understanding the predictability of morphological systems

- the improvement of estuary data

- maintenance/dissemination of ERP2 research and development.

3.2.4 Shoreline management plans (SMPs) and coastal defence strategies

These are of particular relevance to the design process because of the information they provide (see Section 1.4.1 for an introduction to SMPs). Information within SMPs and strategies can include regional sediment transport, forcing conditions (extreme sea levels, waves, sea level rise) and morphological changes (shoreline erosion/accretion, changes in the position of offshore banks or channels). In using this information source, it is important to be aware that the type and level of information available can vary considerably between plans, ranging from qualitative descriptions through to detailed quantification of rates or parameters that are fully based on the results of numerical modelling or direct field studies. In some areas, the information available from SMPs has been further enhanced through the availability of more detailed strategy studies that can consider relevant issues in considerably greater detail (ie improved spatial resolution or location-specific understanding).

3.2.5 Coastal habitat management plans (CHaMPs)

Coastal habitat management plans (see Section 1.4.1 for an introduction) are designed to inform shoreline management plans (SMPs), flood and coastal defence strategies and the planning of maintenance and capital works, to ensure that these plans fulfil the UK Government's obligations under the Habitats and Birds Directives and the Ramsar Convention. They particularly address situations where the conservation of all the existing interests *in situ* is not possible because of natural or quasi-natural changes to shorelines. Their two primary functions are to set the direction for habitat conservation measures to address net change and to assist in accounting for, recording and predicting habitat losses and gains.

CHaMPs can usefully contribute to the baseline understanding required for any management realignment scheme. (Living with the Sea has seven pilot CHaMPs (www[3]), covering coastal areas in eastern and south-eastern England.)

3.2.6 Living with the sea

The good practice guide for habitat restoration *Living with the sea* (English Nature *et al*, 2003) may also provide useful information (www[11]). It is an EC LIFE Natura-funded partnership project, managed by English Nature, and focuses on the impacts of sea level rise and flood and coastal defences on important coastal wildlife areas. Living with the Sea also includes information on how sea level rise could affect the coastline over the next 30–100 years based on the information from the CHaMPs. It covers sustainable and integrated coastal management policies, solutions and best practices.

3.2.7 Local and regional studies

Detailed assessments of physical processes and morphology have been undertaken in many coastal and estuarine locations around the UK. Information from such studies can assist the understanding of historic and contemporary processes and morphology of both: (i) the wider physical setting within which a scheme is set (ie the estuary or coastal sediment cell); and (ii) the environment in the immediate vicinity of the scheme (ie reach- or tidal-flat-scale studies in estuaries and sub-cell or beach-scale studies along the coast). Typical previous work of this nature includes, for example:

- desktop reviews of available literature (eg SCOPAC's sediment transport study, Southern and North Sea Sediment Transport Study (SNSSTS), university theses)
- engineers' reports and supporting technical documentation associated with various flood and coastal defence schemes

- numerical modelling studies associated with proposed flood and coastal defence schemes or developments (eg port expansion, barrage construction, offshore wind farms etc)

- field measurements undertaken in support of the above or as part of ongoing regional or project-specific monitoring programmes or research activities.

Previous studies have included, to varying levels of detail, case study reviews of earlier managed realignment schemes (eg ABP Research & Consultancy, 1998). While some of these reviews usefully include both UK and overseas examples, most have focused on the environmentally driven issue of wetland restoration and consequently have not considered the full range of available managed realignment schemes. The reviews tend to omit those schemes driven principally by physical processes or economic factors. Indeed, current R&D studies are further maintaining the focus on habitat restoration by including case study literature reviews of post-scheme monitoring specifically for the purpose of developing methods for evaluating the ecological and physical "success or failure" of habitat creation within schemes. Although this addresses a key driver for managed realignment it may limit identification of sites that can provide additional habitat benefits within the framework of flood management and private development. Site selection should work to the principles of success and failure criteria (see Section 1.6) and appreciate approaches taken to meet those requirements.

3.2.8 International Navigation Association (PIANC) working groups

PIANC's object is to promote the maintenance and operation of both inland and maritime navigation by fostering progress in the planning, design, construction, improvement, maintenance and operation of inland and maritime waterways and ports and of coastal areas for general use in industrialised and industrialising countries. Facilities for fisheries, sport and recreational navigation are included in PIANC's activities. PIANC undertakes studies through its working groups relating to these areas, especially on wetland restoration and beneficial use of dredged material (see <www.pianc-aipcn.org>). The following studies promoted by PIANC's Environmental Commission (EnviCom) are potentially relevant to managed realignment.

Ecological and Engineering Guidelines for Wetlands Restoration (PIANC EnviCom Working Group 7, 2003)

An international working group that produced best practice guidance for the restoration of wetland for port and harbour managers. The document includes restoration techniques for all types of wetlands and provides guidance on legislation, planning, engineering, ecological functioning and project execution techniques.

Environmental Risk Assessment in Dredging and Dredged Material Disposal (PIANC EnviCom Working Group 10)

This international working group's objective was to provide a consistent and broadly accepted risk-based analysis framework to assess the engineering, environmental, and economic risks of dredging projects and an approach for using these risk estimates in decision-making.

PIANC EnviCom Working Group 2 – Ports & Waterways and Bird Habitats

An international working group that has produced best practice guidance on the management of bird habitats in and around ports and waterways.

3.2.9 A guide to the understanding and management of saltmarshes

This Environment Agency report concentrates on the general management and understanding of saltmarshes, not specifically on managing these habitats for managed realignment. However, the saltmarsh management guide (Halcrow Consulting Engineers, 1994, revised 2001) is a comprehensive guide to saltmarsh management, ecology, amenity and conservation, including estuarine processes, historical change, maintenance and restoration. It contains information on economic values of saltmarshes in relation to coastal defence and other uses. Saltmarsh restoration and regeneration techniques discussed include brushwood fences, sedimentation polders, offshore breakwaters, sediment recharge, vegetation planting and managed realignment. The guide is due to be further updated in 2004.

3.2.10 A guide to managing coastal erosion in beach/dune systems

Scottish Natural Heritage's (2000) report on beach and dune systems is a general introduction to dune processes, the legal framework, strategic approaches to dune management, and management options. It provides sufficient background information to understand general beach/dune management issues. Coastal managers in England and Wales should note that it has been based on Scottish types and legal framework. The focus is on marine erosion rather than wind/recreation/grazing, and therefore should be used as a guide to assist in strengthening and rebuilding sand dunes to reduce the adverse effect of marine erosion (while minimising the risks to people and property behind them).

Figure 3.1   *Sediment trapping and stabilisation of dunes. Guidance on dune management has focused on sediment trapping and stabilisation of dunes, rather than on allowing dune migration inland*

In the appendices, an overview of erosion management options includes planting and management of dune grass and beach recycling and reprofiling. There is also an overview of the monitoring of erosion and change in dune systems. This sets out the reasons for monitoring, suggests what should be measured, considers the measurement methods and frequencies to employ, and offers approaches to analysis. It also discusses management responses, which can be used for initial guidance. Although managed realignment is not formally presented as an option, the advice on "adaptive management" in combination with the softer engineering options is relevant.

A comprehensive study of sand dunes in England and Wales is in progress under the joint Defra/Environment Agency R&D programme.

A guide to the management and restoration of coastal vegetated shingle

This English Nature guide (Doody and Randall, 2003) provides an introduction to the shingle habitat and its attributes and gives guidance on management and restoration. It covers physical and ecological attributes, the importance of shingle structures in flood management, pressures and threats, management and enhancement, monitoring, and funding and legislation of shingle habitats. The English Nature guide deals with aspects of vegetated shingle in addition to other issues for realignment in a general, practical and detailed way. For example, it has diagrams of types of restoration activity and references to practical examples of schemes. Electronic versions of the guides are also available from English Nature (www[11]).

Figure 3.2   *Managed shingle ridges may have artificially high crests and poor vegetation establishment from shingle movement activities*

PART II

4   Communication

Describing the practical methods for a managed realignment scheme can be fairly simple, but explaining the benefits of a scheme to a local community can present a challenge (see Section 1.5).

Experience suggests that timing and style of communication are both important if the audience is not to be antagonised. The process of communicating information, and building up public knowledge, takes some time. The public should be involved from early in the decision-making process through to the development of design, application for consents and licences, and subsequent implementation and monitoring. Members of the community need to understand the steps in the process, feel involved in it and able to discuss their views from the start, and know that their concerns are being listened and responded to. It is also helpful if organisations that support the approach are involved in the communications process. This helps to ensure a wide range of people of differing backgrounds are able to address the different interests in the audience. The acceptance of a scheme by flood and coastal defence authorities, nature conservation regulators, county councils, National Farmers' Union, internal drainage board, Royal Yachting Association, Ramblers Association etc indicates the variety of benefits that the project can deliver and the strength of the scheme's aims and objectives. The lead organisation should ensure that the breadth of issues is addressed sufficiently and, if appropriate, adapt the scheme to provide as many opportunities as possible. If the planning authority is unfamiliar with managed realignment, it may need to be informed and briefed as the project progresses and (potentially) changes.

Larger realignment schemes (say more than 5–10 ha where it is a new concept to the public, or more than 50 ha where previous communication has occurred) will benefit from public display material, with people on hand to explain the project and the process. The longer lead-in periods for larger schemes make it imperative to allow sufficient time for communicating the information, dispelling myths and encouraging ownership of the project. All communication should be handled carefully; where several organisations are involved a communications strategy should be agreed between the parties and adhered to.

Managed realignment schemes need an assessment or proportional consideration of the existing processes of hydrodynamics and geomorphology to explain how the area will change and to show where pressure points already exist. Demonstrating that there is already a problem, caused by natural change, generally makes realignment easier to understand. The main factors in achieving this are hydrodynamics and geomorphology, which define where vulnerable frontages exist and where problems exist in habitat loss.

It is important to explain the benefits of the managed realignment scheme. Scientific information should be reduced to its simplest form and is best supported by using familiar formats for information, particularly maps. Use of GIS can help in both managing the project data and presenting it in an easily understood form. For example, a LiDAR image on one sheet, with hydrodynamics shown as flow increase/decrease/no change can communicate technically complex information effectively. "Before" and "after" views of the site give a clear understanding of the approach being adopted and key impacts (positive and negative) that are expected to occur. Care needs to be taken in communicating the "after" view, as this may change as understanding of the site develops and in response to consultation. Expected changes to social economics, geomorphology, hydrodynamics, habitats and birdlife should not be expressed in detail; general points should be used to communicate the key messages. In particular:

- presentations should have short text with colour graphics
- all detailed information should be appended for reference, and used to back up the scheme for the technical experts to review and ensure robustness
- information must be in the form of documents and deliverables in which the group using has confidence
- it may be necessary to speak directly with those affected to explain the situation, especially to those people who may be unfamiliar with (or even intimidated by) "authority".

Part II – key conclusions

HOW AND WHERE MANAGED REALIGNMENT CAN BE ACHIEVED

The design of a managed realignment scheme is a dynamic process involving several stages, often on an iterative basis to maximise effectiveness.

1   The initial drivers should help to identify the aims and objectives of the scheme. These may be constrained by technical, economic, legal and environmental issues. Criteria to determine the success or failure of the scheme need to be defined at an early stage. Stakeholders should be involved in consultation at an early stage and this process should continue throughout the project.

2   A managed realignment scheme may require a range of consents, permissions and licences. The responsible authorities need to be approached to determine the need for a consent or licence and to identify any specific information needs. The approach to consents and licences needs to be programmed into a managed realignment scheme. Time for consultations, possible revision and resubmission should be included in the programme, together with the costs of applications, which will be incurred whether they are successful or not.

3   The design approach needs to be defined to provide an appropriate level of information for the project. It is important to identify the scale and level of detail in the work to be undertaken in proportion to the size of the scheme and its potential consequences.

PART III
Designing and implementing managed realignment

PART III covers technical issues associated with designing and implementing managed realignment. This chapter is targeted at design disciplines, particularly geomorphologists, modellers and engineers, to assist them in the design, construction and monitoring phases.

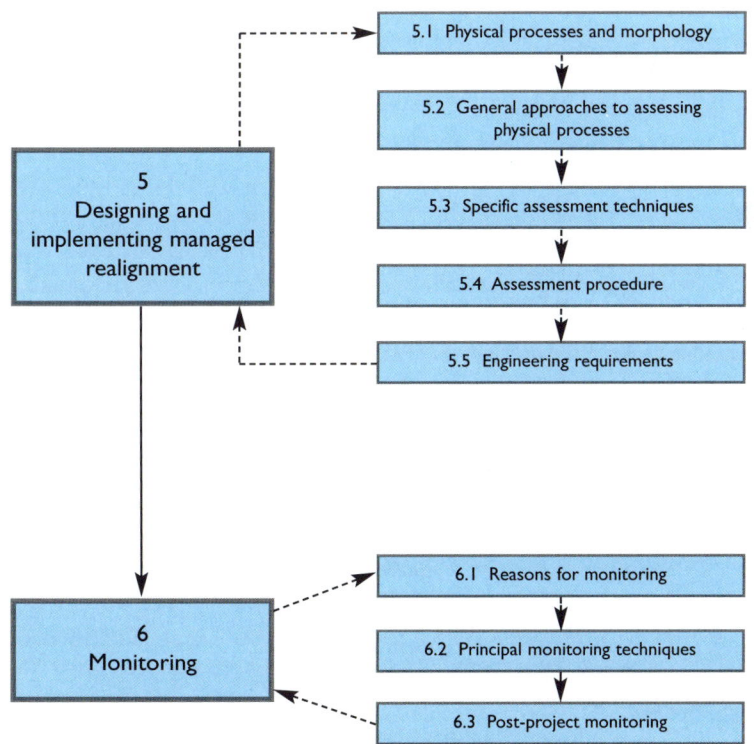

5　Designing and implementing managed realignment

Scheme design is an iterative process that involves:

- identification of a range of initial options or concepts

- development of initial concepts and options, including screening out of unsustainable options

- assessment of impacts or risks associated with the scheme (or schemes) after the initial screening

- further assessment and refinement of the design to reduce the impacts or risks.

To undertake the design process successfully requires specialist contributions covering a range of disciplines, such as physical processes and morphology, engineering, natural and human environment, economics, and health and safety. It is important to consider any assumptions that are made and check data and information integrity (this may require additional survey or monitoring).

5.1　PHYSICAL PROCESSES AND MORPHOLOGY

The design of a successful managed realignment scheme depends largely upon the creation of appropriate physical conditions and morphological responses (see Appendix 3) within the scheme area and adjacent natural environment. Managed realignment schemes require careful assessment of physical processes and morphology, at an appropriate scale, before implementation. These issues should inform the design of any managed realignment scheme for two principal reasons:

- existing physical processes and morphology may influence the design of the scheme

- existing physical processes and morphology may be affected by the design of the scheme, both within the managed realignment site and within the wider coastal or estuarine system (see Box 5.1).

In practice, these two aspects are interrelated. If the initial scheme design is shown to cause measurable effects on the wider system, then the scheme design can be modified to eliminate or lessen the impacts. If they cannot be reduced to acceptable levels this may be a reason not to proceed with realignment. There is no prescribed formula for determining acceptable levels; this will be dependent on the need to accept a change, the degree of change that will occur, the rate of change, the feasibility of implementation, and the likelihood of future adaptation to further natural change.

It is important to present these factors to stakeholders to ascertain their views on what constitutes an acceptable level of change (see Section 1.5). If a scheme has a negative impact on other flood defences where there is a strategy to hold the line, it is unlikely that the scheme will be acceptable. A net loss of a particular habitat might appear unacceptable but may be balanced against improved sustainability and gain in other habitats. The importance of physical processes in the successful design of managed realignment schemes means that these issues are significant also when selecting a site, for example, for development mitigation needs. This is especially the case if the schemes are required to have minimal engineering interventions, such as land level

PART III

change, and/or minimal impacts on sensitive environments. It is also important to consider managed realignment as a tool that can be employed as part of wider flood defence strategies. In such cases it is important to look primarily at creating sustainable and functional systems and to seek to accommodate necessary realignment.

Box 5.1   *Example of scheme design effects on physical processes and morphology*

One of the primary considerations when assessing the potential impacts of a proposed managed realignment scheme, particularly within an estuary, is to determine the changes that will be caused to the tidal prism. Creating new intertidal areas will increase the tidal prism of an estuary, causing more water to enter and leave within the fixed timescale of the tidal cycle. Potentially this could result in increased erosion or deposition at or adjacent to (both updrift and downdrift of) the managed realignment site as tidal velocities are altered. Therefore it is necessary to determine the scale of such changes that might occur as a result of managed realignment.

The effects may include one or more of the following:

- reduction in extreme water levels. This may be a driver behind the managed realignment to mitigate the effects of sea level rise and provide capacity for flood storage
- change to existing intertidal areas, particularly alongshore of the managed realignment site
- the creation of creeks through the existing intertidal area seaward of the managed realignment site. This may result in the change of existing habitat, but may provide an alternative enhanced habitat
- erosion or deposition within deep-water channels, which may affect navigation
- sedimentation not only of the managed realignment site but also of the existing intertidal area outside the site.

For a similar-sized managed realignment scheme, the scale of impact, relative to the existing processes, tends to be less on the open coast than in an estuary (although the scale of effect must be considered case by case). Managed realignment areas may act as potential sediment sinks for material being transported along the coastline. The outflows from such sites may block or cause diversion of the non-cohesive sediments offshore that would otherwise continue to drift alongshore, feeding downdrift beaches. They may also form delta-like deposits ("forebulge") in front of the realigned area that may offer greater protection to the coast locally.

Under the topic of **physical process and morphology**, key parameters in need of consideration are as follows.

(a) Morphological regime (present-day and future evolution)

- Topography (elevation, gradient)
- presence of morphological features (eg creeks, deltas).

(b) Hydrodynamic regime

- Water levels and tidal range
- tidal prism
- tidal current velocities
- tidal asymmetry
- sea level rise
- wave action.

(c) Sediment regime

- Sediment composition
- sediment supply and storage
- sediment transport (bedload and suspended load)
- erosion/deposition patterns.

Each of these physical processes and morphological parameters cannot necessarily be treated in isolation in scheme design or assessment, since many of them interact (see Figures 5.1 and 5.2, Box 5.2 and Appendix 4 for further details and examples). A range of tools and approaches is available to assess these physical processes and morphological parameters to varying levels of detail. Some of these are more applicable to estuarine environments, others to coastal environments. The choice of suitable tools and approaches is an important consideration and for each situation the most appropriate approach(s) will need to be identified, evaluated and proportionally applied. Appendix 2 provides details on the type and appropriateness (or otherwise) of various technical assessment techniques previously used on managed realignment schemes around the UK.

Box 5.2 *Example of feedback interactions between physical parameters and physical response*

> If no earthmoving works are undertaken, the initial elevation of a managed realignment site, relative to the tide, will determine the inundation period of the site and the site tidal prism. These are defined as follows:
>
> - **inundation period (and frequency)** is a prime factor in determining the potential for development of saltmarsh habitats
> - **tidal prism** determines the discharge of water between the site and the estuary and hence the scale of impact on the wider estuary system.
>
> A decision will need to be made whether to undertake the realignment by piping, bank retreat or breach retreat. If culverts are chosen, the capacity of the culverts will control the ingress and egress of tidal water. If breach retreat is chosen, then the size of the breaches, unless artificially reinforced, will be governed by the discharges (and velocities) needed to pass through them.
>
> The hydrodynamics of the site will be influenced by the tidal currents and wave action, the latter being relatively greater for bank retreat schemes than for other forms of realignment because the fetch lengths are greater. The hydrodynamics within a scheme, along with the sediment supply, will play major roles in governing the sedimentation rates. The level of sedimentation within the scheme will in turn play a major role in determining future elevation and tidal prism. These elevations, coupled with tidal prism and wave energy, will be the prime factor influencing the establishment and subsequent development of vegetation (or lack of it) on the site. An additional level of feedback is then initiated, with the vegetation further influencing the degree of wave energy attenuation, sedimentation, land elevation and bed levels within the site.

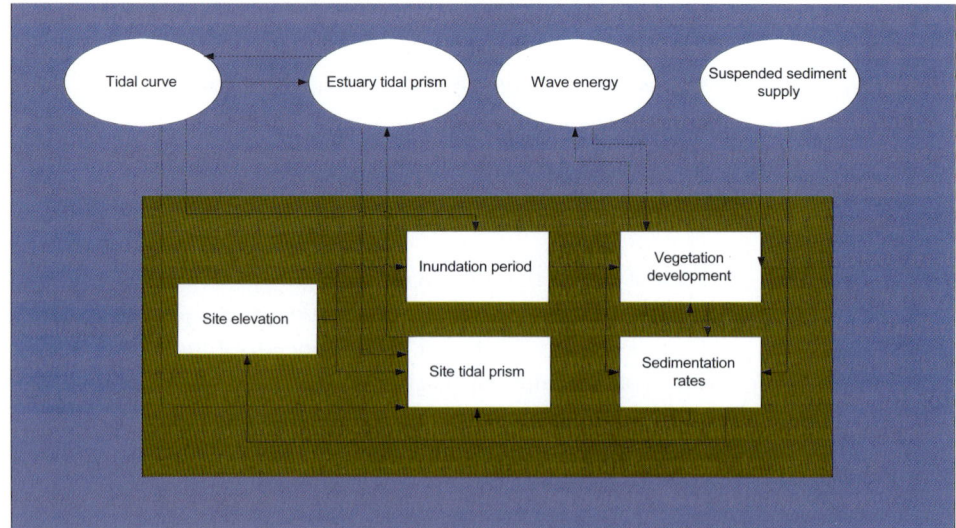

Figure 5.1   *Simplified illustration of hydrodynamic and geomorphological linkages between natural systems (outer box) and managed realignment schemes (inner box) for an estuarine case*

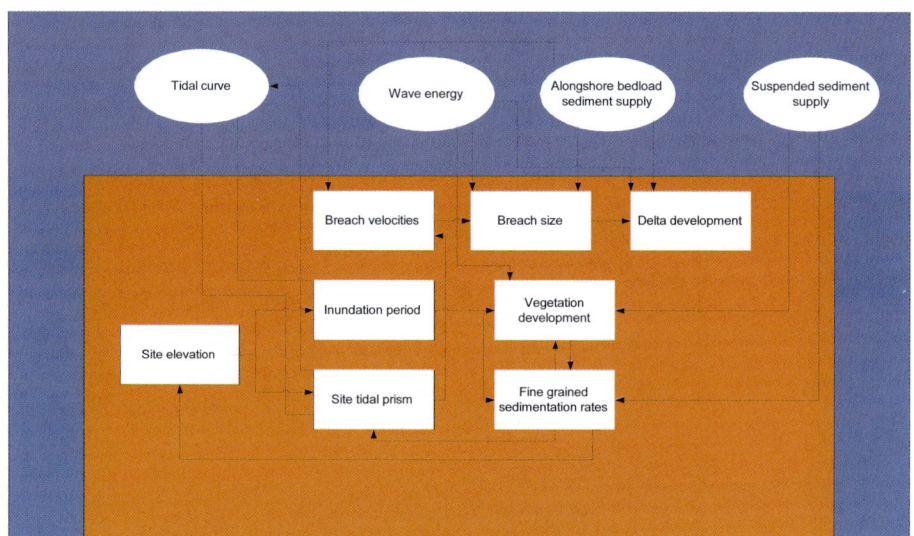

Figure 5.2   *Simplified illustration of hydrodynamic and sediment dynamic linkages between natural systems (outer box) and managed realignment schemes (inner box) for a coastal case*

5.2   GENERAL APPROACHES TO ASSESSESING PHYSICAL PROCESSES

A range of tools exists to assist in the assessment of physical processes associated with managed realignment schemes. These tools can be categorised as desk-based, field-based and numerical model-based. They may be used at different stages of the design process, proportionally, to provide an appropriate degree of certainty in the outcome and may be used in combination to provide a hybrid approach.

5.2.1   Desk-based reviews

Desk-based reviews of scientific and professional literature and/or analysis of existing field measurements can be undertaken to develop a conceptual understanding of the process and morphology. Such approaches are relatively cost-effective and quick to undertake, but are dependent upon the quality and availability of existing literature, data sources and professional expertise. Various empirical and theoretical relationships can be applied to determine the present "condition" of an estuary relative to a theoretical goal, and its role as a source or sink of sediment, or to conceptualise a sediment budget for the open coast. Expert geomorphological assessment (EGA) and historic trend analysis (HTA) will probably be employed (methods for EGA and HTA are being developed further under ERP2; see Section 3.2.3). Desk-based approaches may be the only requirement where the relationship of site area (tidal volume) is small compared with the surrounding estuary or coastline, and the location of the site is not considered to be sensitive. A desk-based study may provide sufficient answers and level of certainty in design to proceed with managed realignment schemes of a few hectares where it is not in a critical location (for the physical or ecological functioning of the system) and/or does not constitute a high proportion of the system in question.

5.2.2 Direct field-based measurements

Direct field-based measurements can be used to capture information on flow velocities, water levels, wave heights, suspended sediment concentrations, sediment transport and erosion/accretion rates. The technical disadvantage of this approach is that the natural system is highly variable and the measurements taken may not be representative of either "normal" or "extreme" conditions. Such measurement should be interpreted carefully and used to advise decisions, not to make them directly. Furthermore, gaining a sufficient spatial spread of measurements over an appropriate period to characterise key interactions would be both difficult and cost-prohibitive. Field-based studies might therefore be targeted on specific locations to address issues, data gaps or uncertainties that have been identified from a broader-level desk-based approach (eg to define sediment grading or shear strength across the managed realignment site or existing foreshore). Data capture might also be needed to support any modelling activity by way of calibration or driving data sets. It may be necessary to capture some data sets even though others already exist. Water level data from Admiralty tide tables, for example, may vary significantly from water levels at the site. Direct field measurement helps address risks where the impacts of concern are identified from a desk-based study regardless of site size and may provide answers for small to medium sites (probably less than 20 ha) where the location of the site is not considered to be sensitive.

5.2.3 Numerical model-based assessments

Numerical model-based (computational) assessments of tidal flows, wave propagation, wave-tidal current interactions, sediment transport pathways and change can be carried out to characterise the existing hydrodynamic, sedimentological and morphological regimes and to determine the extent to which managed realignment schemes and the adjacent open coast or estuarine systems interact through water and sediment exchanges. The disadvantages of this approach are the cost, timescale and input data requirements, and the fact that such models can best characterise only the short- to medium-term processes and morphological changes. Once set up, however, a numerical model can be used to investigate the relative effects of changes in particular processes or morphological conditions on the entire coastal-estuarine system for speific scheme designs, forcing conditions or "what-if" scenarios. Such approaches are important for larger schemes or where detailed scenarios need to be presented irrespective of site size. This approach may also need to be employed where very specific success and failure criteria apply, such as mitigation or compensation for habitat regulation purposes.

Physical modelling has not been used for any managed realignment schemes to date. Physical models can be used to examine specific local features for schemes, such as scour around structures or overtopping of sea defences. For predicting wider environmental impacts, such as changes in flows or sedimentation rates, numerical modelling is generally more suitable.

PART III

The modelling techniques to investigate morphological evolution and changes in physical processes can be divided into top-down models, bottom-up models and hybrid models. This terminology was developed primarily for estuaries under the UK Estuaries Research Programme Phase 1 (EMPHASYS Consortium, 2000a and b) although some of these techniques equally apply to coastal areas.

Top-down models can take two approaches: expert analysis or consideration of regime relationships. Expert analysis synthesises all the available data from various techniques and extrapolates these trends to form a prediction. The regime approach develops relationships between the dimensional features of estuary shape and some measure of tidal flows (eg tidal volume and cross-sectional area). Historic data analysis allows checks to be made against past changes. Top-down models are usually system-wide and extend over long time periods (decades to centuries).

Bottom-up models simulate the physical processes of estuary and coastal areas by solving equations for water movement and sediment transport. These models include most of the commercially available numerical computer-based models for predicting water and sediment movements. Most bottom-up models use finite element or finite difference analytical techniques. The models range in complexity from one-dimensional, through two-dimensional to three-dimensional and can incorporate rectilinear or curvilinear grids. The 3D models are the most sophisticated, enabling flow velocities and suspended sediment transport to be represented within different layers of the water column. For optimum effectiveness, these models rely upon good quality data for input configuration, calibration and validation. Model accuracy can be limited by poor spatial and temporal availability of such data, particularly relating to waves and sediment transport. Bottom-up models are used for short-term predictions (days, weeks, months) and cannot readily be used for long-term predictions because the short-term process or morphological changes are not necessarily cumulative over longer timescales.

Hybrid models are a combination of top-down models and detailed bottom-up process models. The general mode of operation is first to develop a conceptual model and to use the bottom-up model to generate hydrodynamic parameters and sometimes sediment concentrations. These are then used in regime theories, with volumetric or dimensional estimates derived from bathymetric data to explore the position of the system relative to a theoretical equilibrium state. Time series analyses can then be carried out to explore whether the system is moving towards or away from this state. This is the most robust approach presently available, but tools and techniques with respect to predicting long-term evolution continue to be developed and their availability should be reviewed as required.

5.3 SPECIFIC ASSESSMENT TECHNIQUES

5.3.1 Tidal levels and range

On a broad scale, the impact of a scheme on water levels and tidal range within the estuarine or coastal system needs to be carefully assessed since any significant changes can have important consequences for flooding, navigation and the extent of development of natural habitats (see Sections 2.5 and Chapter 5).

Within the site boundaries of a managed realignment scheme, the water levels and periods of inundation represent one of the most important controls in determining the type and extent of formation of various habitats. The frequency/duration of tidal inundation and/or lags in the tidal cycle within the site can determine development of habitats within the scheme. Such considerations are especially important if the scheme objectives require the creation of a specific habitat. This information is also key to the design of any new defences within the scheme so that crest levels are of an adequate height to reduce the risk of overtopping to an acceptable amount.

Predicted tidal levels

If the managed realignment site is close to a location defined by the United Kingdom Hydrographic Office (UKHO) as a primary or secondary port, Admiralty tide tables can provide annual astronomical tidal level predictions. If the site is between standard UKHO ports, linear interpolation can be undertaken to estimate indicative tidal levels in the absence of actual site data. Difficulties may arise in estuarine reaches where ports are far apart or absent. Care should be taken in the application of existing tidal level data, as the morphology of the coast/estuary and the accuracy of measurement may cause significant variation in the quality of data. In addition, actual water levels are susceptible to atmospheric pressure systems that may cause them to differ considerably from those derived from astronomically predicted levels listed in tide tables.

It follows that predicted tidal levels should be interpreted with caution and, where possible, supplemented by field measurements or other data sources at the site. These measurements have usually been short-term (eg weeks), to verify and select synthesised data. Longer periods of measurement generate better data sets, but are appropriate only where warranted by the potential impacts of a project. Captured data from the site is compared with the nearest port to generate more realistic synthetic data statistics. Furthermore, river flows and constrictions in the watercourse can affect water levels locally, making interpolation inaccurate and interpretation complex. In such locations, data may have to be derived from existing river/estuary models or previous study reports (eg feasibility studies for flood embankment construction or maintenance) or from telemetered data (such as those held by the Environment Agency), if they exist.

Recorded water levels

Extreme water levels (comprising both an astronomical component described above, and a meteorologically induced surge component) can be determined through comparison of predicted and recorded levels at a tide gauge or from previous work undertaken by Proudman Oceanographic Laboratory (eg Dixon and Tawn, 1997).

The effects of tidal levels and range on scheme design, or the effects of scheme design on these parameters, can be assessed using:

- mapping within site tidal levels
- professional judgment
- application of numerical models.

Mapping within site tidal levels

The extent of tidal inundation can be assessed by mapping water level predictions on to a digital ground model of the site. This requires detailed coverage of the elevation of the site, which can be obtained by topographic surveying or the use of ground-truthed and filtered LiDAR (light detection and range) data. The accuracy of topographic survey needed depends on the sensitivity of the site to difference in elevation. LiDAR data might not be accurate enough (at ± 15–20 cm) to determine types of vegetation establishment or to calculate site volumes where the existing environment is sensitive to small changes. It is possible to determine whether any new flood embankments will be required to restrict the extent of flooding to confined zones (eg localised bunds around isolated properties), or whether rising backshore topography will naturally constrain the extent of flood zones, using topographic data from the site and adjacent areas.

Professional judgement

In its simplest sense, professional judgement may be used to assess the likely impacts of the scheme on the wider system based on the size of the scheme (in terms of area and tidal prism), the size of the system being considered and the sensitivity of its location, and comparison with similar schemes locally or nationally. While it might be reasonable to assume limited impact of very small schemes (a few hectares) on a very large estuary system or the open coast (thousands or millions of hectares), the impacts of larger schemes within smaller systems, or many smaller schemes implemented in combination, are likely to be harder to assess by professional judgement alone. In such instances, professional judgement probably needs to be informed by a range of other techniques including modelling.

Numerical modelling

In coastal and estuary areas bottom-up numerical models of increasing complexity (1D, 2D or 3D models using rectilinear grids or curvilinear computational grids) can be used to predict the water levels within the scheme and the wider environment (eg entire coastal sediment cell or entire estuary). If the model runs are carried out with and without the proposed scheme in place, then the impacts of the scheme on the existing tidal levels and range can be made. The scale of the scheme and its physical setting will determine the choice and resolution of the hydrodynamic models. These models require calibration against tidal diamonds or measured field data in order for their reliability to be proven.

5.3.2 Tidal prism and discharges

The tidal prism is a measure of the volume of tidal water that would enter and leave a defined area within the time period of a tide, usually measured using MHWS. The tidal prism of a managed realignment site is the volume of water that would inundate a site following scheme implementation and is an important parameter in determining the discharges from a site. This is of relevance to the calculation of velocities through a breach in structural defences, or an inlet in a natural barrier, for example. These velocities can then be used to assess the stability of the breach or inlet, and the potential impacts on the existing intertidal environment, by considering the critical velocities required for erosion, transport and deposition.

In open coast environments, where longshore drift may be an important physical process or where there is a local sediment supply, a breach or inlet may become sealed naturally if the sediment supply is large and the discharge from the site is not competent to remove the sediment. The inward transport of sediment should be assessed in such situations.

Box 5.3   *Example of increased tidal prism considerations*

An increase in tidal prism in relation to that of the area adjacent to the site should be considered. There are no rules for this at present, but changes of more than 5–10 per cent in prism are likely to cause significant changes in morphology.

At one 50 ha site, monitoring has shown no change in flows on most (lower) tides in a narrow creek system, but a 20 per cent increase in peak flow for up to half an hour occurs on the ebb flow of spring tides. This is reflected in the erosion and accretion of the creek bed as a new channel slowly cuts down. It is estimated this process is transporting 5 tonnes of sediment per week, compared with the potential of the existing natural system to erode 500 000 tonnes in a few tides.

Design should aim to spread any impact to meet the background levels of change and avoid noticeable and measurable impact as a net effect. An important principle is that often it is not change itself that poses the problem but the rate of change. The accommodation of gradual change can be incorporated into the design and implementation while the end result may remain the same.

Not all issues identified can be designed out, but many can be addressed by applying knowledge of how systems adapt to change.

The tidal prism of a site can be calculated by subtracting the volume of water over an area at mean low water from the volume of water over the area at mean high water. In most managed realignment sites, it is expected that the site will fully drain at low water unless artificially controlled by sluices or spillway. To calculate the site tidal prism, it is necessary to have a thorough topographic survey of the site, together with sound tidal level data. Digital ground modelling can then be employed to calculate volumes of water below relevant horizontal planes representing the water levels at certain tidal stages (eg MHWS, MLWS). Knowledge of the asymmetry of the tidal curve can again be determined from the UKHO Admiralty tide tables for locations close to primary ports (see Section 5.3.1), or calculated somewhat less precisely based on an assumed symmetrical tide in the absence of such data. This information can be used to determine discharges based on the tidal prism calculations. Information contained within EMPHASYS and Futurecoast can assist in identifying the strength of asymmetry in the tidal curve, but this information does not account for the fact that separate reaches may exhibit differences from the net trend. Hence the site location in relation to alongshore/estuary-wide processes must be considered.

In estuaries, the tidal prism of the site can be an important parameter since it results in the contribution of additional volume to the tidal prism of the estuary as a whole. Unlike a river course, where the mean flow can be considered to be independent of the channel size, the flow within an estuary is dependent on the estuary morphology (Pethick, 1984). That is to say, estuarine morphology dictates the volume of water that can enter and leave an estuary within the period of one tide. If the morphology of the estuary is altered so that the volumetric capacity is either increased (through managed realignment) or decreased (through reclamation) then the volume of water entering the estuary will increase or decrease respectively. Consequently, an increased estuarine tidal prism consequent on managed realignment may influence velocities and morphology (as a result of increased erosion potential) within the estuary (or reaches thereof).

In coastal areas, effects on physical processes caused by the tidal prism of a managed realignment site impacting on the wider system are likely to be less significant than in estuaries (because of the relationship of the site's tidal volume compared with the open sea). However, under a breach retreat scheme, the consideration of site tidal prism and its associated discharges in the area of the breach remains an important consideration in determining impacts on longshore littoral transport; for example, high discharges from the site could potentially flush sediment being transported alongshore in the littoral zone further offshore, or could interrupt sediment supply from updrift to downdrift frontages. Alternatively, low discharges could result in the breach becoming

PART III

sealed by drifting littoral sediments. No models exist that can fully integrate these processes, so a range of approaches will need to be used in a hybrid way.

Various tools can be used to investigate discharge and tidal prism relationships between the managed realignment site and the adjacent coastal or estuarine environment:

- discharge relationship
- English Nature formula
- O'Brien relationship (estuaries only)
- Escoffier curves (coasts only)
- numerical models.

Discharge relationship

The cross-sectional averaged velocity (u, in m/s) through a breach or channel cross-section is given by the relationship:

$$u = Q/A$$

Where:

Q = discharge (m³/s)

A = cross-sectional area of breach or channel (m²).

The discharge can be obtained by considering the tidal prism of the site (Ω, in m³) and the time (t, in seconds) taken for tidal waters to drain from the site, as governed by the tidal asymmetry outside the site and site topography.

Calculated velocities can be compared with critical thresholds for erosion, transport and deposition of various sediment grain sizes to determine whether the breach or channel cross-section will be susceptible to change due to the altered velocities.

Assuming that sufficient bathymetric data exists, this method can also be applied to an entire estuary, through selection of channel cross-sections at appropriate spacings (eg 1 km) along the estuary's length. This can be used to determine whether the increase in averaged velocities downstream of the managed realignment scheme, caused by increases in the discharge through each cross-sectional area, will affect the wider environment (Cooper, 2003). The limitations of this approach are that the calculation considers only an averaged velocity through the channel cross-section and does not represent changes in velocity through the water column. However, this technique can be applied relatively rapidly and cost-effectively to determine the order of magnitude of changes that may be anticipated as a result of the scheme. If these are large, further investigation may be required, but if they are insignificantly small and there are no unresolved issues of contention, then this type of study could suffice.

English Nature formula

An empirical relationship has been defined between log-transformed breach width and tidal prism. This relationship was derived from assessment of unrepaired storm-induced breaches in (clay) flood embankments in estuaries in south-eastern England and can be restated in terms of breach width as an exponential curve whose form is:

$$W = 37.9e^{1.8\times10^{-6}\,TP}$$

Where:

W = breach width (m)

TP = tidal prism (m³) = site area (m²) × depth of water (m) on MHWS.

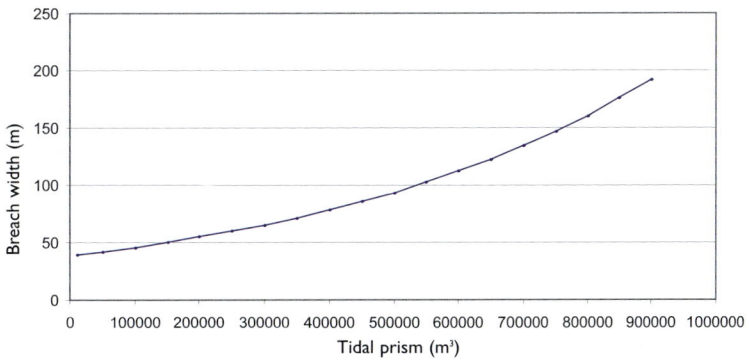

Figure 5.3   *Breach width as a result of bank failure in relation to prism*

The equation demonstrates that breach width is more sensitive to tidal prisms greater than about 5×10^5 m³ (see Figure 5.3). This probably reflects responses in depth (ie scour) where seawalls have failed providing a greater cross-sectional area. The precise design of a breach may not be best achieved by universally applying this formula, but it does provide a useful rule of thumb in muddy estuarine environments with similar characteristics.

> **Health warning**
>
> The English Nature formula was derived based on empirical evidence and thus has no basis in physics and limited explanatory power. The formula was derived following assessment of unrepaired storm breaches in existing flood embankments in Essex and Suffolk covering a limited number and range of tidal prisms of the inundated sites. It was not originally intended for use in the design of artificial breaches within flood embankments; it describes the situation of the existing breaches and does not infer that equilibrium has been reached. Alternative methods are in existence for the design of regime channels based on well-tested physical principles (see for example Selim Yalin and Ferreira da Silva, 2001). A further advantage of these techniques over the English Nature formula is that the width and depth of the regime channel can be determined, rather than just the width using the English Nature formula. Application of regime channel concepts to the design of breaches on managed realignment sites is presently under development and, following application, would be reported in any future update of this guidance document.

O'Brien relationship

The O'Brien (1931) relationship provides a method for assessing the equilibrium cross-sectional area for the entrance to a tidal basin, based upon the volume of water that passes through the cross-section during a tidal cycle, known as the tidal prism. It is an empirically derived measure of inlet stability.

O'Brien ratio = A / Ω

Where:

A = cross-sectional area of estuary mouth at mid-tide level (m²)

Ω = tidal prism (m³).

Based upon empirical relationships observed by Gao and Collins (1994), a theoretical ratio of 1×10^4 equates with a stable inlet. In a similar manner to the discharge relationship, this technique can be used to both influence scheme design (eg identify whether a breach width will be stable) and assess the wider-scale impacts (ie the technique can be applied to an entire estuary to determine whether the scheme could cause changes in overall estuary mouth/channel cross-section stability).

Escoffier curves

Some tidal inlets can shift and migrate along the coastline considerably, whereas others are comparatively fixed and stable. Additionally, a single storm event may force a new inlet through a sand or gravel barrier, or conversely may close an existing inlet. In the classic paper by Escoffier (1940) a method was presented for assessing the stability of tidal inlets on the open coast which recognised the relative importance of the two opposing forces that act in the vicinity of an inlet: (i) the wind, waves and currents that continually transport available sediment towards and into an inlet mouth; and (ii) the flow through the inlet that continually carries sediment either back to the sea or farther into the estuary/bay. The size of a tidal inlet and its permanency are determined by the relative strengths of these opposing forces. Escoffier noted that his method was applicable as a surrogate approach to determining inlet stability where observational records did not enable empirical assessments because of, for example, paucity of historic maps and charts.

The ability (or otherwise) of tidal currents to remove all of the excess sediment that accumulates in the mouth of an inlet is largely dependent on the velocities generated at the time of peak flood and peak ebb tides relative to the thresholds for movement of the appropriate sediment grain sizes. It can be assumed that there is a critical value, V_{cr}, which is the threshold at which sediment in the channel can be mobilised. If the velocities through an inlet exceed this threshold, any newly deposited sediment will be moved (and possibly the existing channel will become eroded), whereas if the threshold is not exceeded the inlet will fill with sediment.

Escoffier (1940) presented a method for determining the mean velocity, V_m, at the time of peak tidal discharge through an inlet when its dimensions, the dimensions of the backing estuary or bay, and the tidal range are known. The value for V_m derived from this formula can be compared with V_{cr} for an appropriate sediment grain size to determine whether or not an inlet will be subject to infilling, but it does not enable an assessment to be made of whether or not the inlet will be stable. The method for undertaking this interpretation makes reference to Figures 5.4–5.7, named Escoffier curves. The horizontal axis of each of these figures represents the inlet cross-sectional flow area, whilst the vertical axis represents velocity. On each graph a horizontal line corresponding to $V_m = V_{cr}$ has been drawn and the intersections of this line with the V_m curve represent inlets whose mouth size is static. From these Escoffier curves, it can be seen that as channel area increases, frictional forces are reduced, but, after peaking, velocities then decrease (this is dictated by the morphology of the backing bay/lagoon and the tidal characteristics of a particular area) and any increases in area simply reduce the velocities through the inlet.

If the line and curve intersect as in Figure 5.4, then two intersections, named "roots", occur. One of these roots is defined by Escoffier as "unstable" and the other as "stable". In Figure 5.5, the curve is tangential to the line, giving rise to an "unstable" root, while in Figure 5.6 no roots exist. A permanent inlet would not be possible under the conditions represented in either Figures 5.5 or 5.6.

This approach is relevant to the design of managed realignment schemes that plan to involve deliberate breaching of natural barrier features along the open coast, particularly in relation to sandy environments. If the permanence of any such breach is to be ensured with minimal maintenance intervention, then the velocity curve must be the two-root type depicted in Figure 5.4 and the breach dimensions should be designed to ensure the inlet falls within the segment BCD on this curve. This will enable the channel to be self-eroding until it reaches the stable condition represented by Point D. If, however, the specific sediment characteristics at the specific location, together with the dimensions of the backing bay/lagoon, mean that the velocity curve is better represented by either a single root curve, or a curve with no roots, then it could be expected that management intervention (ie removal of deposited sediment) will be required in order to keep a breach open.

The Escoffier approach assumed a constant threshold value for V_{cr} (eg for sand a value of 1 m/s was suggested), but Dean (1971) developed this further by calculating an "equilibrium" velocity, V_E, related to tidal prism (Figure 5.7) and calculated based upon O'Brien's (1966) cross-sectional area versus tidal prism relationships. The Escoffier technique has subsequently been encoded by US Army Engineer Waterways Experiment Station (Seabergh and Kraus, 1997). Further discussion of the technique is given in the US Army Corps of Engineers (1984) *Shore protection manual* as well as by Van de Kreeke (1992).

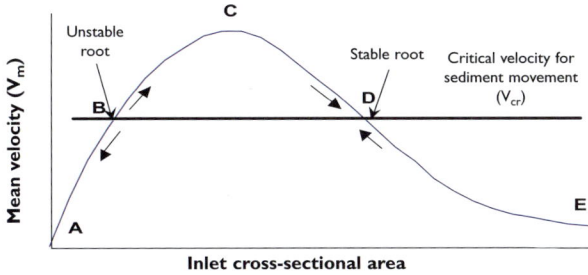

Figure 5.4 Stable and unstable breach sizes. The arrows indicate the evolution of tidal inlets towards or away from the equilibrium condition by erosion or accretion leading to changes in cross-sectional area

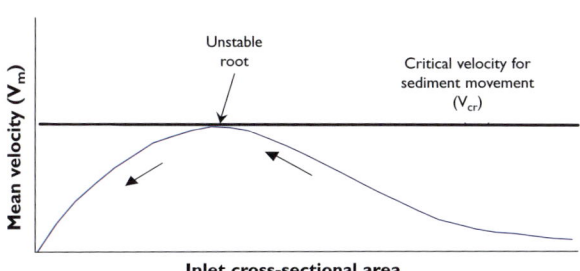

Figure 5.5 Unstable inlet with velocities at or just below the critical value, so that any deposition will remain, causing further decreases in velocity and ultimately closure of the inlet

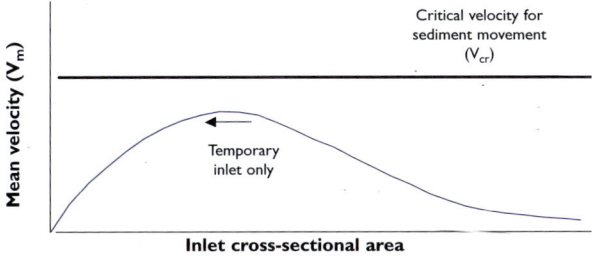

Figure 5.6 Unstable inlet with velocities significantly below the critical value, so that any deposition will remain, causing further decreases in velocity and ultimately closure of the inlet

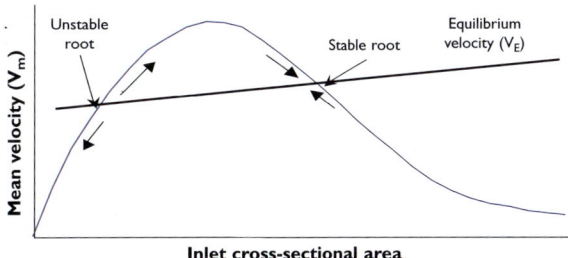

Figure 5.7 Revised version of curve to incorporating an "equilibrium velocity" based on a consideration of tidal prism

Figures 5.4–5.7 *Escoffier curves illustrating the relationship between mean velocity and inlet cross-sectional area*

Application of the Escoffier concept, like the application of all empirical approaches or numerical modelling techniques, should be considered as one of a number of available tools that can be used to assist in assessments of whether a breach in a barrier feature will remain open or reseal. Other valid factors to consider in such an assessment are gained from an appreciation of the wider setting within which the scheme is set. For example, the Escoffier approach does not include an assessment of the availability of river-borne or coastal-erosion-derived sediment supply to the inlet mouth.

Consequently, although the Escoffier results may indicate a theoretical potential for closure of a breached inlet, based upon some hydrodynamic and morphological parameters, if there is no, or little, sediment supply then in practice the inlet may not become closed. This underlines the importance of incorporating a range of appropriate tools in making such assessments.

Numerical models

Numerical modelling would not normally be used to assess tidal prism changes. The application of a numerical modelling approach would incorporate the tidal prism effects in assessments of the changes in current velocities and water levels due to the scheme (see Section 5.3.3).

5.3.3 Tidal current velocities

Tidal current velocities are vectors comprising both current speed and direction, and are an important factor in determining suspended and bedload sediment transport, accretion and erosion rates and locations, both within the managed realignment scheme and farther afield. One of the most important considerations in the design of managed realignment schemes is the velocity of water flow through any breaches that may be created or changes in velocity across the existing intertidal zone due to the discharge of tidal waters from the site. If these velocities are too high then they could lead to significant widening of a breach or erosion of the intertidal area causing feedback. If the discharges are of a very high magnitude, then potentially they also could have wider-reaching effects on the estuarine or coastal system.

Baseline tidal currents will vary in both magnitude and direction as the tidal levels change from low water, through the peak of the flood, to high water and back, through the peak of the ebb, to low water. These currents will also vary through the water column, with near-bed currents being most critical to bedload sediment transport processes and suspended sediment velocities being influenced by the balance between current velocities and sediment settling characteristics (which in turn are dependent on sediment grain size and composition). Managed realignment schemes can also influence general flow directions by redirecting flows on to the site on a flood tide due to the creation of an additional floodable zone or deflecting flows by discharge on the ebb tide.

Figure 5.8   *Potential change to the intertidal zone in front of a realignment area must be considered in the design process*

Tidal currents can be assessed using a range of approaches:

- discharge relationships
- professional judgement
- application of numerical models.

Discharge relationships

Some aspects of tidally induced currents can be investigated by considering discharge relationships (see Section 5.3.2). These relationships can be used in confined channels, but it must be remembered that the calculated velocity yielded from this approach will be a cross-sectionally averaged value. Although this is a useful "first order" indicator of the anticipated magnitude of change, it will not identify peak near-bed velocities that could be critical to the initiation of sediment erosion, transport and deposition processes. For assessment of these more complex processes, numerical modelling techniques will be required.

Professional judgement

In its simplest sense, professional judgement based on identification of the size of the scheme, the size and processes of the adjacent coastal system and a comparison with similar schemes elsewhere may be useful to identify the likely scale of impacts arising from the scheme. However, given the critical importance of water flow through the breach, more confidence can be placed in the results if the professional judgement is informed by other techniques such as empirical calculation or numerical modelling.

Numerical modelling

An assessment of changes in the flow speeds and directions within the wider area of the coastal or estuarine system can be made using numerical models. If the model runs are carried out with and without the proposed scheme in place, then the impacts of the scheme on the existing environment can be estimated. The scale of the scheme and the setting will determine the choice and resolution of the hydrodynamic models. These can vary in complexity from 1D, through 2D to 3D and computations can be performed using rectilinear or curvilinear grids. For changes in flow directions, the more complex 2D or 3D models would be required.

The advantages of this approach are that the model can consider changes over limited spatial and temporal scales, including changes some distance from the scheme and changes at different layers of the water column (if a 3D model is used). The use of 3D models can be beneficial where the system outside the site has strong cross-flows or turbulence associated with, for example, geological control points on the bed/banks or estuary meanders. Significant fluvial discharges into the estuary system, for example, can lead to mixing of saline and freshwater masses, which can result in more complex circulations within the estuary. Consequently, the near-bed velocities can be computed and assessed, rather than just inferred from vertically averaged values.

5.3.4 Tidal asymmetry

For estuarine environments, the consideration of changes in tidal asymmetry caused by managed realignment will indicate whether scheme implementation will have far-reaching consequences. The relative strengths and durations of the flood and ebb tides within an estuary can cause asymmetry in the tidal curve (Friedrichs and Aubrey, 1988), which in turn can influence the net import or export of sediment (although this depends also on the initial availability of sediment supply). For open coasts, asymmetry in tide will influence the flow of water on to and off the site and has a bearing on sedimentation.

Many of these techniques require expert judgement in their interpretation. This is especially the case with the empirical relationships that seek to assess the stability of the system. The exceedance of critical threshold values needs to be considered in the context of the system and sensitivity testing may be required to establish the importance of individual factors.

A range of approaches can be used to quantify these asymmetries in hydrodynamics and sediment transport, including the following.

Empirical relationships

Dronkers derived a ratio (γ) for determining the relative flood or ebb dominance of an estuary (Dronkers, 1998).

$$\gamma = \left(\frac{h+a}{h-a}\right)^2 \cdot \frac{S_{lw}}{S_{hw}}$$

Where:

h = mean hydraulic depth of estuary: $h = a + (V_{lw}/S_{lw})$

a = tidal amplitude of offshore M_2 tidal constituent

S_{lw} = low water surface area

S_{hw} = high water surface area

V_{lw} = low water volume.

Empirical evidence suggests flood dominance occurs when $\gamma > 1$ and ebb dominance when $\gamma < 1$. If $\gamma = 1$, the tidal propagation is symmetrical.

Dronkers showed that if the high water slack period is more protracted than that at low water, then more sediment will be deposited on the upper mudflats at high water than on the lower mudflats at low water. This produces a net landward movement of sediment within the estuary. Conversely, a longer low water slack will lead to the seaward movement of sediment.

Numerical modelling

Output from a numerical modelling exercise can be used to determine the changes in both slack duration and tidal excursion.

Slack duration

Tidal curves can be complex, particularly around the time of slack water. As a consequence, the gradient of slack water is not always representative of the slack duration. Therefore, an alternative approach is required to determine the duration of time when the flow is below a certain threshold, known as v_{slack}. Taking the difference in time between high and low water threshold exceedance values provides a measure for the asymmetry and for the movement of fine sediments, with positive values indicating flood dominance and negative values indicating ebb dominance.

Commonly, the thresholds used for v_{slack} are between 0.1 m/s and 0.2 m/s but will be dependent upon the sediment; the critical value for v_{slack} depends on the value of velocity below which sediment is deposited (see also Section 5.3.6). Thus the value for v_{slack} will vary between locations and should be chosen by considering the size of the sediment that is important in governing morphological responses.

Tidal excursion

Peak velocities on the flood and ebb tides are often used as a first indicator of the preferred direction of movement of coarse-grained sediments. However, this measure takes no account of the duration of such peak velocities. Indeed, it is quite common for a slightly lower velocity on one stage of the tidal curve to prevail for a much longer period than the slightly higher peak value on the opposing stage. To obtain a more representative indicator of the direction of preferential transport of coarser sediment it is necessary to calculate the tidal excursion. This can be achieved by calculating the difference in areas under the curve for the flood and ebb velocities, taking into consideration an appropriate threshold velocity. A positive value indicates flood dominance, and a negative value indicates ebb dominance.

Determining what constitutes fine sediment and coarse sediment (fine sediment refers to mud in suspension, coarse refers to sands and gravels being transported predominantly as bedload), along with the threshold velocity for the initiation of sediment transport, will vary depending on the system being considered. Again, expert judgement is needed, supplemented where necessary by field measurement, to identify which sediment type is important in determining the morphological response.

Figure 5.9 presents example output of assessments of both slack duration and tidal excursion for a particular estuarine location, as derived from a numerical model. These baseline conditions for a specific location within an estuary can be compared against post-implementation modelling output to ascertain the scale of change in these parameters.

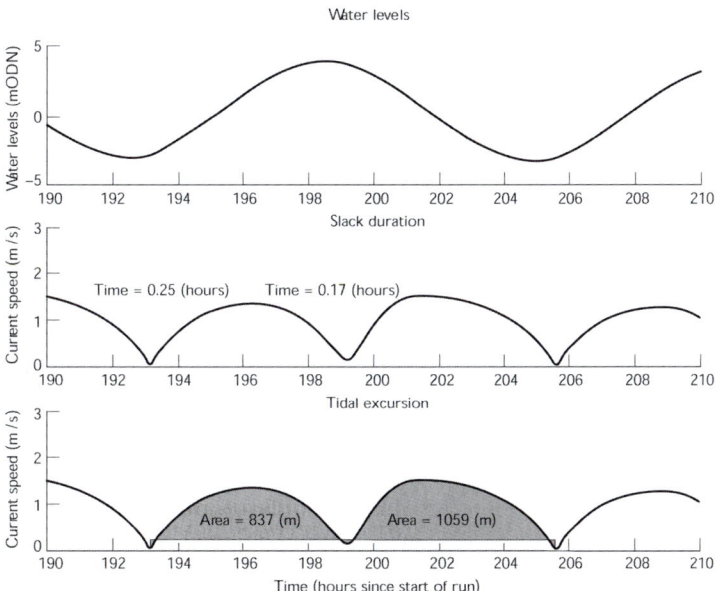

Figure 5.9   *Assessment of slack duration (middle plot) and tidal excursion (bottom plot) for given water level conditions (top plot)*

5.3.5   Waves

Waves are important processes to consider when designing managed realignment schemes. The wave energy incident on an existing flood defence structure (seawall, flood embankment) or natural feature (dune or gravel barrier) may be a driving force behind the decision to realign. Additionally, once this decision has been made, it is important to understand how wave processes may affect the design of a scheme, or vice versa. In particular, a "breach scheme" located in an area where waves are experienced would need to anticipate the diffraction of waves through the breach. Larger managed realignment schemes may also need to consider the effect of locally generated wind-waves within the scheme area.

Waves propagating into a realignment scheme, or waves generated within it, are of importance when predicting the rates and locations of erosion and deposition of sediment within the site. This factor links to colonisation and subsequent development by floral and faunal communities. An assessment of wave energy is also critical to the design of new sea defences within the scheme, eg design of crest levels to reduce the risk of wave overtopping, and any need for protective armouring of embankments.

In general terms, the principal wave processes that need to be considered on open coast managed realignment schemes differ from those that require consideration on schemes within more sheltered estuarine environments. This is because open coast sites are generally located where it is important to have an understanding of offshore wave generation and transformation of offshore waves towards the shoreline. In managed realignment schemes located towards the mouths of large estuaries, such processes will similarly need consideration, but for many other estuarine schemes it is locally derived (within scheme) wind-wave effects that are more relevant.

Offshore wave climate

When considering managed realignment schemes along the open coast, it is first necessary to understand the offshore wave climate. This is characterised by "deep water waves" – defined as waves typically within 15–20 m or more of water, which are not limited by water depths. Such waves can be measured using directional wave recorders, or modelled using hindcasting techniques. The offshore wave climate is often characterised by wave height, period and direction, although a fuller description of the sea state can be given by the frequency spectrum, which gives the distribution of wave energy as a function of frequency.

The offshore wave climate comprises both wind-sea and swell. Wind-sea refers to wave energy generated within a limited area (up to about 1000 km radius) by winds occurring relatively recently (within the preceding day or so). Swell refers to waves that have moved well away from the area in which they were generated. In practice, the distinction between wind-sea and swell is not always clear or relevant. Around the UK, wind-sea usually has larger wave heights than swell, but swell has higher wave periods and tends to arrive predominantly from directions where the fetches are very long (1000 km or more).

Using available offshore data, it is possible to create frequency tables depicting wave height against direction or period, and to produce wave roses that are a pictorial representation of wave heights from different directions. It is also possible to apply theoretical probability distributions, such as Weibull (see Simm, 1996), to offshore data and predict extreme conditions under different return period events. However, the reliability of this approach depends on the length and accuracy of the data record to which the probabilistic distribution is fitted. A commonly applied rule of thumb is that reasonably reliable predictions can be made for return periods of up to three times the length of the data record. For example, if 10 years' worth of data is available, it may be reasonable to use this data to determine return periods of up to 1 in 30 years. In practice, unfortunately, inappropriately long return periods are commonly calculated from relatively short data sets and the results produced can become cast in stone. As the length of the data set improves, so does statistical fit to the data, as does any estimation of return periods beyond the limits of the data record.

Wave transformation processes

As offshore (or deep-water) waves propagate towards the shoreline they become transformed by various processes. These result in the inshore wave climate being markedly different from the offshore wave climate. The *Beach management manual* (Simm, 1996) presents a summary of the most important wave transformation processes in shallow water, and the most commonly applied assessment methods (more detailed information is contained within CIRIA/CUR, 1991). The roughness measure applied to a wave transformation should take account of the surface over which it is travelling. Field evidence (Möller *et al*, 1996, 1999, 2001, 2002) has shown that the roughness introduced by saltmarsh surfaces (from vegetation, creeks, mud-mounds etc) significantly reduces wave heights in *average* spring tide inundation conditions. No data is yet available for extreme inundation events such as may occur during a storm surge.

Wave attenuation

A reduction in the magnitude or energy of a wave is termed wave attenuation (see A5 for wave attenuation formulae). This occurs through either the movement of material (ie using the kinetic energy of the waves to move objects such as sediment or vegetation), or friction (energy is converted to heat energy and water motion slows down as a result). Attenuation occurs when waves arrive at a shore where there is vegetation (Asano *et al*, 1992), such as on saltmarshes, or a morphological roughness element (caused by surface topography or sediment type), or fluid mud (Whitehouse *et al*, 2000) or percolation into the voids between the sediment causing a "damping" effect on the wave energy.

Wave attention is important to consider when designing a managed realignment scheme because of its potential for reducing wave heights, thereby enabling vegetation to remain *in situ* and/or decreasing the impact on to the flood defence structures needed to implement the scheme. Indeed, the wave attenuation effect of mudflats and saltmarshes serves as a first line of defence for defence structures. A saltmarsh in front of a defence structure reduces routine maintenance costs for the defence line and increases the defence lifespan under "normal" tidal/wave conditions. The marsh's attenuation capacity during extreme events is unclear, however. It is conceivable that, once a given inundation threshold is exceeded (eg during a storm surge), the surface roughness does not significantly affect wave propagation. If this is the case, then the actual height of a defence might not be lower than that designed using a usual wave transformation approach (to the new alignment position).

It has been suggested that a 6 m-wide saltmarsh can halve the required seawall height and reduce the cost of building the defence by one-third. A wave height reduction of approximately 40 per cent over an 80 m-wide saltmarsh from shoaling and breaking processes and from frictional losses is supported by early laboratory-based work undertaken by HR Wallingford, based on a representative morphology of saltmarsh with fronting cant (cliff) and using monochromatic waves (individual waves of uniform height and frequency) (Brampton, 1992). The fact that field conditions are characterised by complex wave spectra, and the addition of more recent work, suggests modification of this analysis is required, so caution should be applied when using it.

Recent initial comparisons of wave attenuation patterns observed in the field with the scale model data (Möller, pers comm) suggest that the scale model results in overestimates of attenuation immediately landward of the saltmarsh cliff. The field measurements on the east coast of England show that during inundation depths of about 1 m saltmarshes attenuate waves non-linearly with distance, rapid attenuation occurring in the first tens of metres from the leading (seaward) edge of the vegetation.

PART III

Very little or no further attenuation occurs beyond about 70 m inland of the marsh edge. If a cliff is present at the marsh edge, wave heights increase across the cliff face before experiencing attenuation.

Past sea defence cost calculations need to be reviewed in the light of this new evidence. It is more likely that the presence of saltmarsh reduces both the maintenance costs of sea defence and the risk of overtopping/damage under average conditions, rather than reducing the design height of a defence under higher return period events (notwithstanding that a realignment landward may allow a lower defence if there is greater wave transformation across the new, wider intertidal area).

The value of attenuation should not be underestimated, however. Recent monitoring in the Wash by the Environment Agency has measured mudflats reducing incoming wave heights by up to 36 per cent, with the higher saltmarsh further reducing the wave height by up to 91 per cent of the original incoming height. Overall energy was reduced by up to 56 per cent by mudflats, and 97 per cent by the time the wave had propagated to the upper marsh (Coastal Geomorphological Partnership, 2001). A wave tank study in America by Fonseca and Cahalan (1992) indicated that when the seagrass height was approximately equal to water depth, an energy reduction of about 40 per cent was observed within 1 m width of seagrass (they considered this *relatively* independently of species type and vegetation density). This effectiveness was reduced, however, as water depth exceeded "canopy" height, ie when the water was above the top of the vegetation.

The factors influencing attenuation have been explored in an Environment Agency R&D project (W5B-022) and are outlined in Figure 5.10.

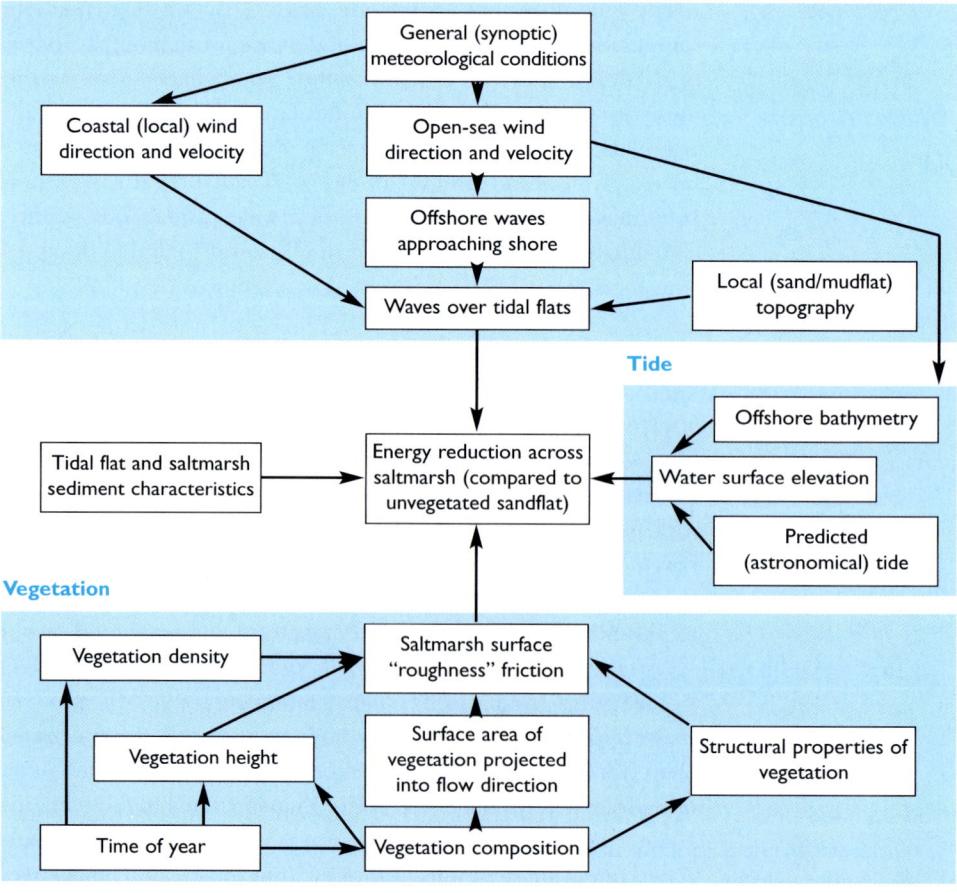

Figure 5.10 *Factors influencing wave attenuation over saltmarsh surfaces (Spencer et al, 2003)*

The Environment Agency report (Spencer *et al*, 2003) includes the following material of relevance to managed realignment.

Saltmarshes exist within a larger-scale coastal geomorphological context. Water depths and incident wave conditions play an important role in determining the relative reduction in wave energy across the marsh surface. Knowledge of larger-scale geomorphology (such as offshore topography, cross-shore and longshore slopes, and large-scale creek/or estuary flows) may need to be taken into account to calculate the expected water depths and incident wave energy at the marsh edge (Spencer *et al*, 2003).

Short-term wave energy attenuation may not be sustainable in the long term. Saltmarshes can be relatively effective at attenuating waves over individual inundations or a series of inundations (see Section 5.3.5). This, however, may be due to energy dissipation by the movement of sediment off the marsh surface by erosion, which itself reduces the future wave attenuation capacity of the saltmarsh. The tolerance thresholds of saltmarshes to incoming waves are yet to be determined.

Saltmarshes differ significantly in terms of species composition and this may have implications for energy dissipation potential. Different species present different "canopy" profiles to the incoming waves and have differing degrees of physical flexibility (see Appendix 5).

The initial success of any saltmarsh re-creation depends on the balance of hydrodynamic and sedimentological influences at the site. Fluidisation of the bed and water viscosity can also have some attenuation effects. The shear stresses in the pore water (without significant movement of the pore water) and an associated build-up in the pore water pressure can fluidise a bed and transfer some wave energy to the bed. This tends to occur in sediments of low permeability (ie mud) in relatively unconsolidated conditions. The presence of sediment in the water will increase both the bulk density and the bulk viscosity of the suspension (Delo, 1988). The viscosity of fluid mud is at least several hundred times larger than that of clear water. The damping of waves over a fluid mud bed, showing dissipation of wave energy, has been observed in flumes and in the laboratory (Whitehouse *et al*, 2000). Salinity gradients can also occur in estuaries, thus causing different water densities that may also cause a damping affect on waves (if the density is increased).

Refraction

Refraction is the change in *direction* of wave propagation. This generally arises due to changes in wave propagation velocity when waves propagate in varying water depth. In this situation, the direction of wave incidence relative to the beach inclines towards the direction normal to the local depth contours. Refraction can also occur independently of water depth changes when the presence of a varying current field alters wave propagation velocity. This issue is particularly relevant in areas dominated by strong tidal flows.

Shoaling

Shoaling is a change in wave *height* due to waves propagating into different water depths and is a separate effect to changes in wave height that may be a consequence of changes in wave direction due to refraction.

Diffraction

Diffraction is propagating waves that impinge on obstacles such as piles, breakwaters, headlands and islands and interacting with these structures. The resulting wave field around these structures or features generally shows a marked change relative to the undisturbed wave field. This process is important in managed realignment schemes where waves are able to propagate into the site through any breach(es) in the fronting defence.

Reflection

Waves propagating on to a structure can reflect off it, combining with the incoming waves to increase orbital bed velocities that can cause problems in terms of scour at the base of existing or realigned defences, for example.

Breaking

Wave breaking is a non-linear surf-zone process that can be dependent on either depth or wave steepness. Breaking occurs when the maximum wave height becomes limited by either criterion. The way that energy is distributed to the seabed/intertidal sediments varies according to the way the wave breaks and may lead to the construction (raising) or destruction (lowering) of the bed/intertidal area (Komar, 1998; Allen and Pye, 1992).

Friction

Seabed and internal friction dissipates energy as waves propagate towards the shoreline. Generally, the longer the distance over which wave is affected by a sloping seabed/intertidal the more it is altered, changing its shape and the way energy is dissipated. The degree of dissipation is dependent on, among other factors, the surface roughness of the seabed. In vegetated intertidal areas, this surface roughness plays an important role in attenuating energy that may otherwise impact on backing defence structures or natural features. This issue is relevant to managed realignment schemes intended to lessen the pressure on the coastal system through the lengthening of an intertidal profile. Further information on this topic is contained within Appendix 5.

5.3.6 Investigation of waves

In estuarine and coastal areas, a range of approaches can be adopted to investigate waves:

- professional judgement
- empirical equations
- numerical models.

Professional judgement

Professional judgement can be exercised to determine the relative importance of wave processes to any particular managed realignment scheme based upon results from a desktop assessment of wave conditions recorded within the wider environment, or through specific knowledge of the conditions experienced near to the site.

Experience of typical wave generation and propagation conditions within certain physical environments will be of assistance in exercising this professional judgement.

Offshore wave data can be purchased from the Meteorological Office (modelled offshore data) or obtained from:

- existing studies (eg previous defence design, shoreline management plans or Futurecoast)

- strategic monitoring campaigns (such as on the Anglian and south coast or the Defra-funded Wavenet programme)

- field measurements.

In terms of general wave transformation processes from offshore, it may be expected that wave exposure at managed realignment sites along the open coast or near the mouths of large estuaries may be greater than for sites located up-estuary, where tidal processes become more dominant. Equally, at locations more exposed to wave activity, it may be expected that wave-induced sediment transport is important. Factors that can influence both initial wave generation and subsequent wave propagation from offshore towards a managed realignment site can include:

- fetch length and both wind-sea and swell exposure (these factors will determine the magnitude of potential wave energy incident on a site)

- offshore and nearshore bathymetry (ie the presence of banks and seabed gradients that will induce refraction, or the levels of the seabed relative to water depths which will influence shoaling and breaking)

- undulations in the shoreline planform (ie headlands or structures that may cause wave diffraction)

- the length and nature of the intertidal zone, which acts as a buffer zone in dissipating wave energy propagating to the shoreline (eg shingle foreshores dissipate energy through percolation, vegetated intertidal flats create a surface roughness that dissipates energy through leaf blades baffling the water motion).

The above factors can be used to inform an assessment of the relative importance of wave processes to the managed realignment scheme. Based upon experience of previous schemes, it may also influence decisions about whether processes such as wave attenuation due to bed roughness (ie saltmarsh vegetation), percolation in shingle barriers, or diffraction through breaches etc are in need of further consideration using empirical and/or numerical modelling approaches.

On the open coast modelling approaches are well developed. In estuarine environments a desktop study of existing wave effects can be undertaken to determine the "limit of influence" of wave activity. This can be based on known data from various reaches of the estuary, from an indicative assessment of offshore wave directions and the exposure of the main estuary channel to these directions, and identification of any zones of differences in the sediment composition – for example, coarser sediment close to the estuary mouth suggests that wave influences may extend to these areas.

Empirical equations

To describe the full sea state of waves, the *Beach management manual* (Simm, 1996) presents a variety of semi-empirical wave spectra, each with a specific range of applicability. The two most widely used are the spectrum described by Pierson and Moskowitz (P-M) (Pierson and Moskowitz, 1964) and JONSWAP (Hasselmann *et al*, 1973). The P-M spectrum represents a fully developed sea in deep water, while the JONSWAP spectrum represents fetch-limited sea states. CIRIA/CUR (1991) describes the modification to the JONSWAP spectrum for use in a wave generation area of limited water depth. It also describes how several different wave height and period parameters may be derived from the spectrum.

PART III

The most relevant of these parameters are:

$$H_s = 4m_0^{1/2}$$

$$T_m = (m_0/m_2)^{1/2}$$

Where:

H_s = significant wave height, which is approximately equal to the average height of the highest one-third of the waves

T_m = average wave period

m_0 and m_2 are moments of the wave energy density spectrum.

Both deep-water and shallow-water waves can be estimated using design curves (see Figures 5.11 and 5.12) that are based on an effective fetch, a wind speed (although some methods prefer the theoretically more correct wind stress) and wind duration, yielding a significant wave height and a wave period. The wave height is limited either by fetch (when the duration is long) or by the duration (when the fetch is long).

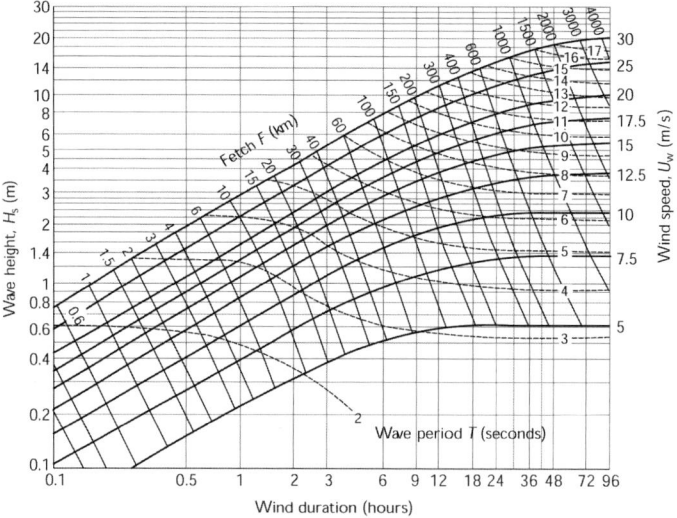

Figure 5.11   *Deep-water forecasting diagram for a standard wind field*

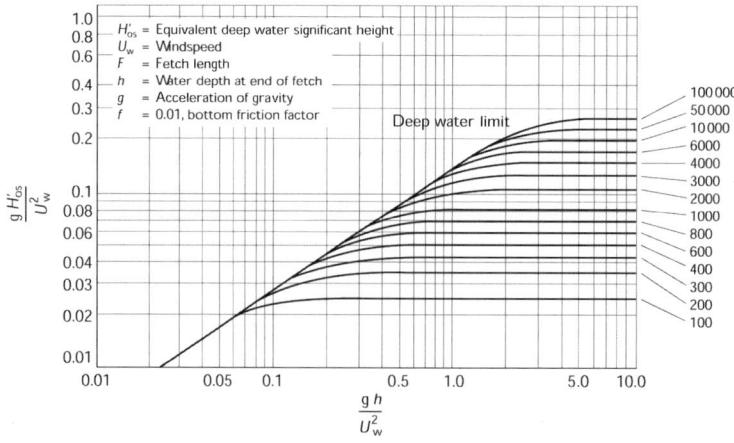

Figure 5.12   *Shallow-water wave forecasting diagram for a standard wind field*

Alternatively, an assessment of locally generated wave heights can be made based on the methods outlined by McConnell (1998), which involve consideration of the fetch length, wind speed and the depth of water (h) relative to the wave length (L). Although this approximation method does not take into account the frequency of winds from different directions or the influence of the seabed bathymetry, it can be a useful indicator of the magnitude of exposure, so it can be used as a method to discern those schemes where wave conditions may need to be more robustly assessed through numerical modelling of wave transformation.

Previous work has been undertaken on the transformation of wave energy across intertidal surfaces from the processes of friction (eg caused by the hydraulic roughness created by saltmarsh vegetation), percolation (most relevant to beach profiles composed of coarse clastic sediments), shoaling and viscous damping. Various empirical and theoretical formulae exist (see Appendix 5).

Wave overtopping is an important issue to consider should there be the requirement associated with a managed realignment scheme for the construction of a new set-back line of defences or cross-bank/counter-wall defences to limit the extent of tidal inundation to defined boundaries. This requirement often does not exist where the managed realignment site is surrounded by rising topography. Overtopping of coastal defences generally occurs when:

- water levels alone exceed the crest level of the defence
- waves impact the structure and run up the face of the structure and over the crest
- when water levels are near the crest of the structure and waves break at, or near, the crest.

"Green-water" overtopping is defined as being when a continuous sheet of water passes over the defence crest. Other forms of overtopping include splash and spray overtopping. Splash overtopping is caused where waves break on the face of a defence structure and the resulting water droplets are carried over the defence crest, either through their own momentum, or as a consequence of a strong onshore wind. Spray overtopping is caused by strong onshore winds acting on wave crests immediately offshore of a defence. Methods for calculating overtopping discharges under various wave and water level conditions are well established for the types of seawalls, normally encountered in managed realignment schemes, and have previously been fully described in the Environment Agency's *Overtopping manual* (Besley, 1999).

The input parameters required for these overtopping calculations include wave conditions at the base of the defence structure. In the case of a realigned defence line, it is important to account for the processes of wave transformation across the increased intertidal width, incorporating processes of friction, percolation, shoaling and viscous damping. Failing to account for such wave-damping effects could result in the over-design of a defence structure. However, exposure of the site before vegetation becomes established, for example, needs to be taken into consideration too. This may require the inclusion of some temporary or sacrificial works.

Numerical models

In coastal and estuarine areas, bottom-up numerical models can be used to: (i) transform offshore waves inshore and, if applicable, into an estuary; and (ii) create locally wind-driven waves within the realignment site. Such models can be used to predict the wave conditions both within the scheme and elsewhere in the wider environment and they can be run with different water level conditions as input. Representative boundary wave conditions (heights, periods, directions) can be derived from frequency and extremes analyses of an available time-series record from offshore sites. For modelling of wind-generated waves, an assessment of meteorological records can be used to define input wind conditions.

PART III

The Futurecoast study presented a useful application of the modelling transformation of offshore wave conditions for 68 inshore locations around the coastline of England and Wales (not including estuaries). These inshore wave data were derived from analysis (using a numerical simulation of wave energy transfer based upon backtracking wave ray modelling) of resultant wave data (wind-wave and swell-wave) obtained from the UK Met Office at 24 offshore locations, comprising three-hourly records for the period 1991–2000. The transformation of data from offshore to inshore was derived from linear wave theory, which enabled inclusion of the processes of refraction, breaking, shoaling, wave-wave interaction and dissipation of the wave energy due to bottom friction. Dual-parameter analysis was undertaken on the transformed inshore time series to produce frequency distribution summaries and wave roses.

If wave model runs are carried out with and without the proposed scheme in place, then the impacts of the scheme on the existing wave regime can be made. The choice and resolution of the wave models will be determined by the scale of the scheme, its physical setting and the input requirements (eg offshore or inshore wave data, and the availability, quality and quantity of data to represent the bathymetry of the seabed etc). For example, it will be necessary to determine the relative importance of processes of diffraction, refraction, reflection, shoaling, wind growth, bottom friction, depth-induced breaking, white-capping, and both wave-current and wave-wave interactions prior to defining the most appropriate modelling techniques. The relative importance of waves will vary for sites in different locations. Generally, waves are a more important issue to consider in coastal environments than in more protected estuarine environments, but large schemes in estuaries can still be subject to internally generated wind-waves.

5.3.6 Sediment dynamics and morphological response

Sediment dynamics, and resulting morphological responses, are governed largely by the availability and composition of sediment and the hydrodynamic forcing conditions that are operating. Consequently, an understanding of the physical environment in which the managed realignment scheme is set is of critical importance when undertaking design and assessment.

Schemes within estuaries tend to be dominated by cohesive sediments and tidal action, and typically an understanding of intertidal flats and saltmarsh processes will be of importance, especially if the aim of the scheme is to create or re-create these particular habitat types (including reedbed as a transitional saltmarsh species). In contrast, schemes along the open coast may be more influenced by non-cohesive sediments and wave action (or combined wave and tidal action). In such environments, a managed realignment scheme may relate to the landward movement or breaching of a structural defence (eg flood embankment or seawall) or a natural feature, such as a sand dune or shingle ridge, and an understanding of these landforms and their formation and evolution processes becomes important. Further information relating to a range of these coastal and estuarine landforms can be found in Appendix 2. It is important to consider the information presented in Appendix 2 when designing a managed realignment scheme in these environments, or when aiming to re-create specific habitat types, as previously described in Section 2.5.2.

More information of relevance to the processes and landform responses along the open coast can be found in CIRIA's *Beach management manual* (Simm, 1996), while in estuarine environments the Defra/Environment Agency EMPHASYS study output provides similar supporting information. Defra's Futurecoast study contains a geomorphology manual, which provides information of use in understanding the formation and evolution of a wide range of landforms. Classic textbooks on such processes include Pethick (1984), Carter (1990) and Komar (1998).

Sediment dynamics covers aspects of sediment supply, transport, erosion and accretion. These aspects play an important role in determining conditions within the site, as well as the impacts of the site on the wider environment. For example, sediment erosion might lead to the development of creek systems and the expansion of breaches, while deposition might lead to the natural raising of land levels within a site or, in a breach realignment scheme, to the creation of flood or ebb tide deltas at the entrance to the site. Outside the scheme, morphological impacts might include changes in the cross-sectional shape of the intertidal mudflat creeks or estuary channel in response to increases in flow discharge (tidal prism) due to the scheme.

Figure 5.13   *Sediment accretion can occur within managed realignment sites, raising land levels and developing new creek networks. Unless sediments de-water they can be easily eroded*

Dealing with cohesive sediments can present difficulty in predicting sediment suspension, resuspension, and settling due to the nature of the processes. Flocculation is the process that binds together several silt/clay particles to form larger particles by electrostatic forces (a similar process can occur through mucus secreted on to sediments or through faecal deposition from marine fauna). It is a process that is critical to determining the potential settling of sediment and thus sediment accretion in managed realignment sites. A range of references deals with this process (see the summary in Whitehouse *et al*, 2000).

Within the site, a major consideration is the expected location and rate of sedimentation or erosion, since this determines bed level changes, which in turn are important in governing the hydrodynamic exposure conditions and, in certain schemes, habitat type. The type of sediment in the area of the managed realignment influences the magnitude and rate of morphological change. For example, vegetated cohesive sediments may have considerable resistance to erosion, while non-cohesive silts and fine sands may be more readily eroded. Such issues are also help determine the rate and equilibrium depth of scour or the development of creek networks. The erosion of sediments from a managed realignment site may affect water quality, especially if the sediments are contaminated above background levels and are released into the wider environment. Erosion in certain areas is especially likely in breach realignment schemes, where the exchange of water flow between the scheme and the wider coastal or estuarine system is confined and consequently inflow and egress velocities can be higher. The transport of eroded material away from the site boundaries can affect the wider environment. For example, such erosion, transport and subsequent deposition could have detrimental effects on nearby navigation channels or (shell)fisheries; conversely it could help in the accretion of other intertidal areas or managed realignment sites farther up the estuary.

In coastal areas, managed realignment schemes have the potential to interrupt, or be affected by, longshore drift patterns, which can have affect the sediment supply to areas along the coast as well as the stability of the breach itself.

Examples include:

- discharges from a site creating a hydraulic barrier as a "jet" of water flowing out of the site can interrupt longshore transport, creating accumulation updrift and reducing supply to downdrift frontages

- discharges from a site flushing littoral sediment into the nearshore zone where it may be transported offshore or recirculated in the nearshore zone

- newly created spits or tidal deltas

- set-back alignments creating changes in alongshore transport potential that may lead to increased accumulation within sink areas

- high rates of alongshore sediment supply sealing breaches or inlets.

In such instances an assessment of the discharge potential in and out of the managed realignment site (see Sections 5.3.2 and 5.3.3) needs to be compared with the predicted longshore transport rate of material (and local redistribution). It is also important to consider the width, and sediment nature, of the intertidal where sand dunes are being realigned. Changes in beach sediment and morphology will affect the amount of sand material that can be blown by the wind and may even present a physical barrier to landward sand transport. The resultant change from realignment may have positive or negative implications for the coast dependent on where, and to what extent sediments accumulate or deplete.

5.3.7 Investigation of sediment dynamics

In estuarine and coastal areas, a range of approaches can be adopted to investigate sediment dynamics:

- professional judgement
- empirical equations
- sediment modelling.

Professional judgement

This can be based on evidence of sediment behaviour from similar schemes. For example, measured accretion rates on other realignment schemes in similar environments, or knowledge of general accretion rates within adjacent intertidal areas, can assist in estimation of anticipated accretion rates within a managed realignment scheme. A historical trend analysis and expert geomorphological assessment may help in this assessment. Many realignment schemes will be changing the environment back towards earlier configurations.

Empirical equations

If the velocities due to tidal flow, wave action or the combined action of both waves and tides within a managed realignment site or across the adjacent coastal or estuarine area are known, then a number of empirical relationships can be used to determine the likelihood of erosion, transport and deposition of sediment. Such relationships are usually some variation of that first proposed by Shields in 1936 (the Shields curve; Rouse, 1939).

Soulsby (1998) extends the classic work of Shields to establish the threshold of motion for sediments beneath waves and currents (see Figure 5.14).

Where:

$$\theta_{cr} = \frac{\tau_{cr}}{g(\rho_s - \rho)d}$$

And:

$$D_* = \left[\frac{g(s-1)}{\upsilon^2}\right]^{1/3} d$$

With:

θ_{cr} = threshold Shields parameter

τ_{cr} = threshold bed shear stress

ρ_s = grain density

ρ = water density

d = grain diameter

g = acceleration due to gravity

$s = \rho_{s-\rho}$

v = kinematic viscosity of water.

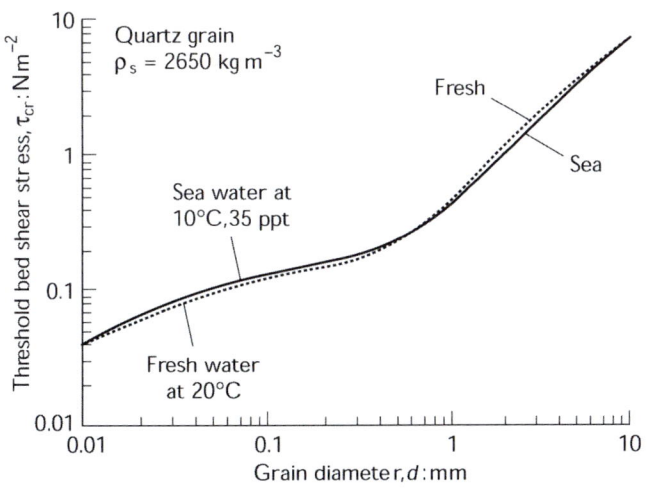

Figure 5.14   *Threshold of motion for sediments beneath waves and currents (Soulsby, 1997)*

Similar approaches are available to determine the threshold of motion under waves alone. Komar and Miller (1975), for example, relate the threshold of movement for different sizes of quartz particles to orbital velocity and wave period (Figure 5.15). Similarly, the US Army Corps of Engineers' *Shore protection manual* (USACE, 1984) provides graphs to enable longshore transport to be estimated as a function of wave breaker height and breaker angle. Chapter 2.4 of CIRIA's *Beach management manual* (Simm, 1996) presents the detail of various approaches to investigating sediment dynamics. A number of standard textbooks provide further information on such processes (eg Dyer, 1986).

The application of these empirical formulae can be used as a basis for assessing the likelihood for the erosion, transportation and deposition of both suspended and bedload sediments within managed realignment sites, and the potential for longshore sediment transport along the open coast.

PART III

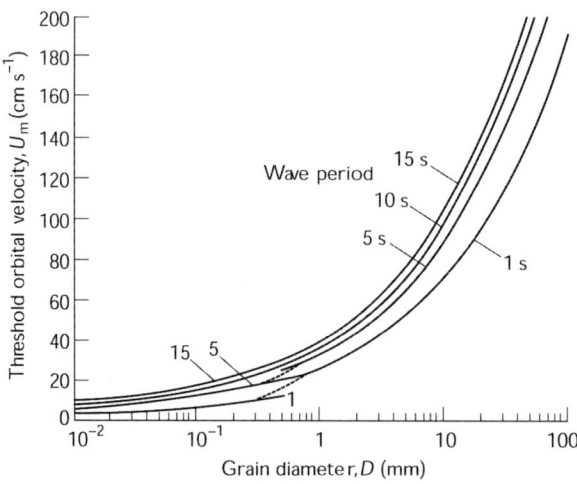

Figure 5.15 *Near-bed orbital velocity for the threshold of sediment movement under waves of different period (Komar and Miller, 1976; taken from Dyer, 1986)*

Sediment modelling

Although certain modelling tools are in existence, much research and development effort is still required to improve model representation of sedimentary processes and morphological responses in the marine environment.

A constraint on sediment modelling is the prediction of the effective (wave-related and current-related) bed roughness as a function of hydrodynamic and sediment conditions. At present, these parameters are used for calibration, but ideally the bed roughness should be determined as an integral part of the computational method. This would require the ability to predict the shape and dimensions of the bedforms through different computational time-steps. Although such "morphodynamic" models exist, they are at an early stage of development.

The available morphodynamic models incorporate calculations of the transport of various fractions of sediment (shingle, sand or silt) within the flow and/or combined wave-current calculations. Calculations of both suspended load transport and bedload transport are performed and the bed level is updated during each time step of the flow computations, taking into account the exchange with the suspended sediment vertical profile and the gradient of the bedload transport. Through identification of areas of erosion and sedimentation, this enables the effect of morphological changes to be assessed. The models tend to incorporate a morphological scaling factor, enabling changes simulated over computation timescales (eg days, weeks or months) to be scaled to represent changes over periods of months or even years.

In general, less confidence can be placed in the results of the morphodynamic modelling than in the results of the hydrodynamic modelling. This arises because (i) the hydrodynamic model can be more accurately calibrated and (ii) the number of assumptions and possible sources of error are greater for the morphodynamic model. Also, reliable sediment transport field data are still very scarce, particularly under storm conditions with breaking waves, and inter-comparison tests of model predictions often show quite diverse results.

Notwithstanding the above limitations, sediment and/or morphodynamic modelling can be undertaken with existing tools to assess the following factors:

- predicted changes in potential wave-driven longshore drift along the open coast due to realignment in the plan form of the coastline. This can be of use for determining the wider-scale impacts of coastal realignment schemes on downdrift, and potentially updrift, frontages

- predicted changes in suspended sediment concentrations and pathways within the wider environment due to managed realignment schemes within estuaries or other cohesive sedimentary environments

- predicted input of sediment/redistribution of sediment within or erosion of sediment from a realignment site due to both bedload transport and suspended transport mechanisms

- the potential of a newly created embayment to act as a net sink for material. In a littoral sense this may trap material from longshore drift, whereas an estuary might generally act as a quiescent area for the settling of suspended material.

5.4 ASSESSMENT PROCEDURE

5.4.1 Appropriate level of technical assessment

To enable the most appropriate approach to be adopted in the consideration of physical processes during the design and assessment of managed realignment schemes, both the applicability of the potential techniques and the proportionality of the approach will need to be considered.

Management of data

When starting an assessment it is important to consider management of the data and information and to ensure that the technology used is appropriate. This may include examining issues such as:

- data collection technology
- the extent to which the technology is "future-proof"
- storage technology, including the appropriateness of database systems (including meta-data), whether simple systems or more comprehensive GIS-based systems are used
- data visualisation
- back-up/disaster-recovery technology
- data dissemination.

The same technology, if possible, should be used for both data storage and distribution, especially if remote access (eg Internet) is envisaged.

Selection of techniques to investigate physical processes

Many techniques are available to investigate the physical processes and morphological issues relevant to the design or assessment of a managed realignment scheme (see Section 5.3). However, these techniques should not all be adopted for the design or assessment of managed realignment schemes in all cases. Instead, the methods employed should be appropriate and proportional to the specific characteristics of, and issues associated with, the scheme and need to be agreed at the outset of a project between all interested parties.

Proportionality

The level of assessment should generally be proportional to the size and location of the scheme in the context of the coastal or estuarine system within which it is set and the type and contentiousness of issues raised by statutory (and non-statutory) consultees. A small managed realignment scheme of a few hectares, at a non-sensitive location, within a large estuarine system, for example, may not necessarily require detailed numerical modelling, even if the estuary has statutory nature conservation designations. However, previous experience has shown that overriding stakeholder interest, particularly from conservation or flood defence groups, can often lead to a disproportionate amount of assessment being required.

Selection of the most appropriate technique or (more likely) combination of techniques will depend upon factors related to the scheme, as summarised below:

- the physical environment in which the scheme is set (ie the coast or estuary), which can influence the complexity of the physical processes and morphological issues. This should include an assessment of the geology and sediment availability in the surrounding environment

- the location of the scheme within the coastal or estuarine system (eg near the head or mouth of an estuary, or at the updrift or downdrift limits of a coastal sediment cell)

- the size of the scheme

- whether the defence to be realigned is natural or man-made

- the aims and objectives of the scheme and associated key design requirements

- the need to have a more prescribed outcome (eg in terms of the creation of specific habitats or the incorporation of specific landscape features)

- how contentious the scheme is (eg in terms of environmental, landscape, political and economic aspects).

The above factors need to be considered at the outset of a project in order to identify the potential risks associated with a particular scheme and to assist in the identification of the appropriate tools to undertake the necessary assessments. The relevance of some of these factors is described further below.

Physical environment

Coastal and estuarine environments vary in terms of the dominant physical processes in operation, so different issues may need to be considered. For example, in simplistic terms many coastal areas are dominated by wave action, whereas most estuarine environments are less influenced by waves than by tidal action.

Water movements within estuaries are more complex than on coasts because of the enclosed channel morphology and the influence of freshwater and saline inputs. The unidirectional freshwater flow from upstream interacts with the bi-directional saline tidal flow from the downstream reaches, resulting in the mixing (to varying degrees) of these two bodies of water. This process sets up residual currents, which can be directed up or across the estuary. These water movements are further complicated by the presence of surface waves. As well as waves formed within the estuary, waves can also reflect across an estuary and be generated externally (ie offshore) and propagate into the estuary. The complexity of water movements is reflected in the sediment transport pathways within the system. In comparison with coastal environments, longshore littoral transport rates are usually less significant within estuaries. Instead, increases in the estuarine tidal prism due to the scheme are often a more important consideration, since these can lead to changes in tidal velocities and directions and changes to the estuary's morphology.

In a coastal environment, the potential interruption of longshore sediment transport pathways by a managed realignment scheme is an important issue, because this can cause downdrift erosion (should a sediment sink be created as a result of the realigned planform position of the shoreline) or can lead to the blockage of the entrance to a managed breach. In the example of a newly created inlet through a managed breach in an open coast barrier or defence structure, high rates of sediment exchange between the wider system and the scheme can lead to the formation of flood or ebb deltas at the inlet mouth.

It is clear that the choice of assessment technique needs to ensure the capability to sufficiently assess the appropriate physical processes that operate in different physical environments. Whilst it can be expected that some physical process considerations will be similar between estuarine and coastal environments (ie for any breach scheme, site tidal prism will be an important consideration irrespective of its physical setting), other processes are more relevant to one environment only. This means that some techniques that are applicable to managed realignment schemes in estuarine environments may not necessarily also be relevant in coastal environments, and vice-versa. The principal differences are that managed realignment schemes in coastal environments will, in addition to the assessment of issues common to both environments, generally require assessments of:

- non-cohesive sediment transport (and within-site sedimentation rates)
- formation of ebb and flood tide deltas
- longshore drift.

Scheme size and location

The required level of assessment of physical processes and morphological parameters will also differ depending on the size and location of the managed Realignment scheme. Within an estuary, for example, it is the size and location that may be important in determining scheme impacts on the wider environment. Larger schemes are likely to require a greater degree of assessment, since the potential impacts of the scheme are likely to be of greater potential significance to the existing "baseline" conditions. Schemes in upper reaches of estuaries may exert relatively more influence on water levels than schemes (of the same size) in lower reaches.

Scheme complexity and contentiousness

For particularly large or complex schemes, the multi-faceted nature of coastal and estuarine processes and the limitations of various techniques often makes it advisable to use a range of techniques and synthesise results to develop a conceptual understanding of the processes and geomorphological functioning of the area being considered. This approach is explained within the document "A guide to prediction of morphological change within estuarine systems" (EMPHASYS consortium, 2000b). This approach might use a range of techniques to assess many of the parameters described in Section 5, including the application of various empirical equations and numerical modelling. In the case of less complex schemes, a lower level of assessment may suffice and may be based on the application of a few empirical relationships and a comparison with similar schemes elsewhere.

The above applies similarly for contentious schemes, which may or may not be unduly complex.

A prime driver in determining the degree of assessment required of physical processes may be the environmental interest in the vicinity of the scheme. In many cases, these environmental interests will be protected under a range of national and international conservation designations including special protection areas (SPA), special areas of conservation (SAC), Ramsar sites, sites of special scientific interest (SSSI), ASSI (in Northern Ireland) and areas of outstanding natural beauty (AONB).

The enforcement of these designations by the relevant conservation authority, plus the requirements of other stakeholder groups, will often define the scale of physical and natural impact assessment required for a given scheme (see Section 2.5.1). This will largely be governed by the need to demonstrate no significant adverse impacts on key geographical areas or physical features (processes, landforms or habitats). Key physical process issues will often include assessments of the potential for erosion of designated areas such as saltmarsh, mudflat, beach or dune habitats.

The specificity of such issues will often require some of the more robust and accepted techniques described in Section 5.3 to be applied in place of professional judgement or assessments of changes in gross properties of estuarine or coastal systems.

Time should be allowed for consultation and discussion and implementation of the consent procedures. In general, the more contentious the project, the more time will be needed, as more information needs to be considered and more techniques will have to be applied.

Scheme aim and purposes

In terms of the evolution of the managed realignment site itself, the degree of process assessment necessary may vary depending on the aim and purposes (see Section 1.1) that it is hoped it will deliver. A scheme being used as mitigation for habitat that will be lost as a result of a proposed development, for example, may need to deliver specific habitat requirements. This will require accurate predictions of bed levels and sedimentation rates as well as a detailed knowledge of the elevation preferences for individual vegetation types. Conversely, a scheme implemented with the objective of reducing economic expenditure on maintenance of existing defences may be less concerned with these aspects, since the type of habitat formed may be secondary to other issues and considered as adding value.

Other relevant issues

The level and type of assessment also depends upon other issues relevant to key stakeholders such as planning authority, statutory consultees, shoreline and seabed users and local residents. These can often be identified through consultation/scoping exercises undertaken at an early stage of the study. Impacts on, for example, socio-economic and local community parameters, navigation, recreational use, marinas, ports, fisheries, access and water-using industry may all be of concern (see Section 2.5.1 on environmental impact assessment).

5.4.2 Applying proportionality

To ensure that the techniques provide an appropriate and proportionate level of assessment, a staged approach can be used (see Figure 5.16). Initially this should involve collation of information (Stage 1) and application of a range of empirical/ theoretical relationships (Stage 2). If these approaches are considered to have satisfactorily addressed any scheme-related process issues, as defined in Chapters 2 and 3, it is suggested that the scheme could move on to synthesising the results and assessing the impacts. If high levels of risk or sensitivity are associated with the scheme then numerical modelling may be required (Stage 3). This is essentially a risk-based approach to determine the appropriate techniques to apply.

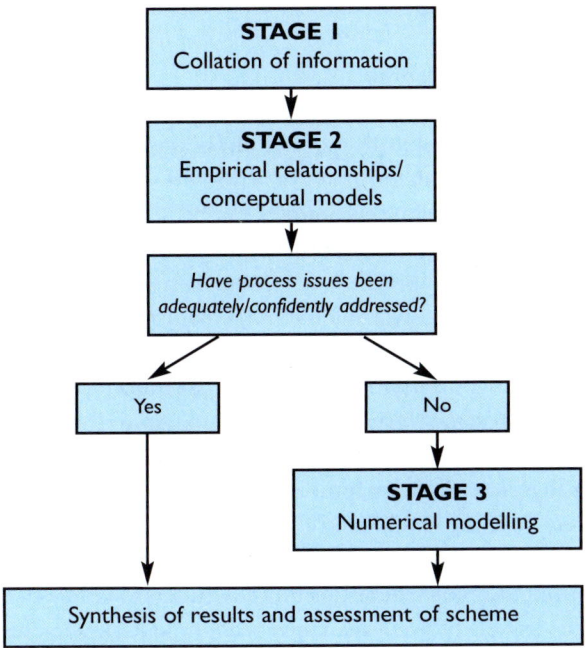

Figure 5.16   *Simplified staged approach to assessing physical impacts of managed realignment*

Each of the three stages is described below.

Stage 1: Collation of information

Initially, information should be collated to identify:

● the size and location of the scheme within the wider-scale setting

● the range of issues to be considered (as a minimum these are likely to include investigations of the hydrodynamic and sediment regime and corresponding morphological responses and the effect on any designated environment)

● the techniques available to address these issues (which may be different for coastal and estuarine environments).

Much of the information required to make the desk-based assessment will be available from existing sources although it may be time-consuming to gather (see Sections 2.3, 2.4 and 2.5).

Stage 2: Empirical relationships/conceptual models

If the size, location, complexity and/or contentiousness of the scheme, together with the specificity of the issues, suggest that the risks associated with the scheme are likely to be low then expert judgement or empirical relationships can be used as a first-order assessment stage.

A first-order assessment of the magnitude of changes in physical processes and morphology attributable to the scheme can be based upon the assessment of various empirical relationships under the baseline conditions, and the conditions with the scheme in place. Much of the information required to feed into these assessments can be derived from existing information sources such as Futurecoast, Admiralty charts and tide tables, and the EMPHASYS database. Dependent on the magnitude of changes in various relationships, it can be decided whether the gross impacts of the scheme are so small as not to pose any significant risk, or whether residual risk and uncertainty remains unacceptably high and also whether more specific assessments are required.

Stage 3: Numerical modelling

Where extremely detailed assessment of physical processes is required, or risks are perceived, or have been determined to be high through empirical relationships in Stage 2, numerical modelling techniques can be used to inform the assessment. It should be noted that numerical modelling alone would not determine the significance of the location and magnitude of any scheme-related changes in the hydrodynamic or sediment regime. Rather, professional judgement needs to be used in interpreting the results from numerical modelling in the context of changes to the baseline regime. Numerical modelling approaches require suitable input data for the exercise to be worthwhile, and in some areas such data are sparse and may require collection. These aspects could add to the cost, and significantly increase the timescale, of proposed managed realignment if data capture is required. However, the absence of suitable data could severely limit the confidence in model output.

Typical assessment requirements, technically suited to either coastal or estuarine environments under different managed realignment implementation scenarios, are presented in Table 5.1. Expert geomorphological assessment, historic trend analysis, desktop study and modelling will use these techniques as necessary.

Table 5.1 *Example of information requirements and approaches for managed realignment sites in coastal and estuarine environments*

	Estuary	Coast
Breaches	Site tidal prism calculations; changes in gross estuary properties, discharge relationships; numerical modelling	Site tidal prism; Escoffier curves for inlet stability; sediment budget assessments (sediment supply/storage, changes in longshore transport potential, inlet bypassing mechanisms, tidal delta formation); sand/shingle beach evolution modelling (planform/cross-shore).
Bank realignment	Site tidal prism; changes in gross estuary properties, discharge relationships; numerical modelling; wave attenuation caused by lengthened profile	Sediment budget assessments (sediment supply/storage, changes in longshore transport potential, potential for creation of sediment sink); wave attenuation caused by lengthened profile; sand/shingle beach evolution modelling (planform/cross-shore).
Tidal exchange system	Controlled tidal prism, changes in gross estuary properties, discharge relationships, wave attenuation over vegetation	Controlled tidal prism; changes in longshore transport potential, tidal delta formation, sand/shingle beach evolution modelling (planform).

Assessing risk

To determine which of the above techniques is (or are) the most appropriate, a generic approach can be taken, using expert professional judgement to identify the level of risk. Table 5.2 shows examples of how the degree of risk might be determined by placing the site in context of the main physical parameters existing outside the site, such as the tidal prism and sediment characteristics, and the management risk and/or level of controversy that a scheme is likely to attract. Such an approach can be applied to as many criteria as necessary but should relate to the primary aim of the managed realignment.

Table 5.2   *Examples of scales of physical and coastal or estuarine management issues for managed realignment sites*

	Relative magnitude of issue	Example situations	
		Estuary	**Coast**
Physical issues	High	Large site prism with respect to estuary prism	Large scheme adjacent to unconsolidated coast with high rates of alongshore transport
	Medium	Moderate site prism with respect to estuary prism	Moderate scheme adjacent to unconsolidated coast Large scheme adjacent to hard rock coast with minimal alongshore transport
	Low	Small site prism with respect to estuary prism	Small to moderate scheme adjacent to unconsolidated coast Moderate to large scheme adjacent to hard rock coast with minimal alongshore transport
Coastal or estuarine management issues	High	Presence of internationally designated nature conservation/earth heritage/landscape sites. Presence of many other shoreline and seabed users and uses, eg shellfisheries, ports, flood and coastal defence, disposal sites.	
	Medium	Presence of some designated nature conservation/earth heritage/landscape sites. Presence of some other shoreline and seabed users and uses, eg shellfisheries, ports, flood and coastal defence, disposal sites.	
	Low	No designated nature conservation/earth heritage/landscape sites. No other shoreline and seabed users and uses affected, for example, shellfisheries, ports, flood and coastal defence, disposal sites.	

Having identified the key parameters, processes and issues and assessed their risks, it is possible to categorise the level of risk in broad terms (see Table 5.3) and determine the appropriate approach. This table combines both the physical parameters of the site (in context) and the perceived risks for management of the site. Importantly, this aims to link the actual physical risks to the perceived risks of the scheme in a balanced manner so that no single factor dominates the decision about the technique(s) required to be used for scheme assessment.

Table 5.3   *Suggested approach to determining the level of assessment of physical processes required for managed realignment sites*

		Physical issues		
		High	**Medium**	**Low**
Management issues	High	A	A	B
	Medium	A	B	C
	Low	B	C	C

Level C sites are likely to be the least complex physically and environmentally and to pose low risks. Use of empirical/theoretical relationships and professional judgement alone may suffice. Level B sites potentially have greater environmental impacts and management risks and are likely to require Level C techniques plus at least some numerical modelling. This modelling may be at a relatively low level of detail to help determine the degree of change and act as supporting information to the professional judgement. Level A sites pose the greatest potential risk of environmental impacts and may have significant uncertainties attached to the predictions of these impacts. Such sites are likely to require Level B techniques with detailed numerical modelling undertaken to increase confidence in the results. The environmental impact assessment conducted in the early stages of the project plays an important role in identifying environmental characteristics and impacts (see Section 2.5).

5.5 ENGINEERING REQUIREMENTS

This section provides information on some of the engineering requirements to consider when designing a managed realignment and is derived from scheme experience. The intention is to highlight issues that might require particular attention or may differ in some way from other strategic options.

5.5.1 Site boundaries

Once an area for implementing managed realignment has been identified, the boundaries of the site are defined taking into account a number of factors.

It is preferable, both economically and environmentally, to retreat to naturally occurring high ground, to create a shoreline that obviates the need for a new, engineered, defence. This may be possible, for example, on a narrow coastal floodplain, but may not be on a wide estuarial plain. In some areas, a combination of retreat to high ground and a new defence (eg counter-walls) may be appropriate.

The presence of any built development within or close to the proposed realignment site can be a significant issue. In some cases, local defences or ring banks may overcome the problem of a small area of built development in a realignment site. Land ownership boundaries and the degree of interest and participation by individual landowners may also constrain the site boundaries.

5.5.2 Realigned defence

Substantial hard defence structures are usually required only where there is a particularly aggressive wave climate, where the width of the new intertidal area is small and/or where there are particular economic assets that restrict the extent of realignment. Substantial defences are more likely to be required in open coast/estuary mouth or urban situations than in estuarine and rural locations. The rural nature of most existing managed realignment sites means that, to date, defences have mainly been earth embankments with suitable scour protection, if required. The newly realigned defence will need to be designed in accordance with usual practice for the design of tidal defences, although certain aspects are worth highlighting.

Figure 5.17   *Realigned defence should be designed and constructed to withstand incident forces, which may include modification to existing secondary defences*

Realignment of defences may require new wave modelling/assessment for engineering design purposes in exposed locations or where the site is large. The larger managed realignment sites may need to account for waves generated locally (within the sites) in addition to waves generated offshore (outside the site). Such waves would normally be forecast by the use of prediction charts that combine the wave climate prevailing at the site with the available fetch lengths, local fetch lengths being important.

The crest height of the realigned defence may be determined by overtopping analysis, to limit the overtopping of the defence to permissible values. This will take account of the type of crest and back slope of the defence for combinations of water levels and waves during an extreme event, usually derived from the nearest joint probability analysis point. This analysis might be adjusted to take account of any local water level assessment (measurements) or fetch conditions. The assessment of waves propagating to the new defence line should account for the wave transformation across the new intertidal width and wave attenuation where appropriate. An allowance for future sea level rise also should be included.

Attenuation by vegetation or surface roughness (see Section 5.3.5 and Appendix 5 for further information) would not usually be applied in the design of the crest level of a new realigned defence, but it may be relevant to the width, armouring or maintenance of the defence. Where water depths are less than about 1 m (because of relatively high land level to water level, or where a tidal exchange system has been designed to raise levels sufficiently prior to breach or bank realignment) the added attenuation of mid- to high-level saltmarsh vegetation might be taken into account in the design. If the new intertidal area is likely to be covered by only annual species, however, attenuation from vegetation should be considered ineffective, because it will not exist in winter months, and only the sediment properties and wave transformation process will be applicable.

The maximum wave acting on the realigned defence during an extreme event is determined from joint probability analysis, subject to the wave transformation/ attenuation appropriate for the new intertidal area and allowances for water levels and sea level rise. This is assessed in a similar manner to determination of the defence crest height. Guidance as to the type and design of scour protection to resist the wave attack can be found in McConnell (1998). The majority of realignments to date have not employed armouring of the banks. No specific guidance appears to be available on the maximum wave that can be sustained by a grassed slope. Based on engineering judgement and experience, it is suggested that scour protection should be considered where the maximum wave exceeds 0.3 m in height at the defence.

The wave attenuation properties of the new intertidal area, ignoring sea level rise and deep-water conditions, will be lowest immediately after it has been established – ie after existing unsustainable vegetation has died off and before new salt-tolerant vegetation has become established or potential accretion has taken place. This might need to be taken into account in the defence design. In some cases, sediment is pushed up in front of the realigned defence, which avoids the need to provide any scour protection. This arrangement should be applied with care as it will add to costs and the material is likely to be eroded easily (for example, during a stormy series of high spring tides), leaving the realigned defence exposed to attack. The use of a sediment wedge may nevertheless provide a sacrificial erosion zone that provides some temporary protection to the flood defence. It might be applied to the new alignment or to the remains of the original defence. This is only likely to be needed where some time is needed for vegetation to become established on the site (providing attenuation) and for accretion to take place (naturally raising the level). This is essentially buying some time for the defence while the site establishes itself and modifies before it becomes exposed to the force of the natural processes.

PART III

Ground investigation will be required to determine both the conditions on which the realigned defence will be founded and the parameters of the material from which the defence may be constructed. Many of the sites proposed within estuaries may be alluvium, probably silty in nature, and with high moisture content. They may also have lenses of coarser material (sand and gravel) from past channels or geological features and variable conditions where creeks within a reclaimed area have been infilled. Cone penetration test or odometer tests can help in determining the stability of the surface and understanding the nature of the material.

An earth embankment could be constructed from material won from a local borrow pit or any other earthworks on the site (such as creeks, land drainage or environmental enhancement works), thereby minimising the importation of material to site and offering additional environmental opportunities. This approach will help not only to minimise costs but also to reduce the local environmental impact during construction. Consideration should be given to conditioning the material before construction or to employing construction methods that enable pore water pressures to dissipate. This may be achieved by following construction with a period of bank stabilisation (and reshaping) prior to breaching/retreating/tidal exchange through the existing defences.

Bank retreat is ideal for creating an integral part of the natural environment if there is certainty in the hydrodynamics and associated impacts. This is an expensive approach relative to breach realignment. The Construction (Design and Management) Regulations (CDM) place a duty on those undertaking construction to ensure construction workers are not put at risk, and bank retreat may increase the risks above other approaches for realignment. Working across tidal cycles in intertidal areas needs care to avoid health and safety and environmental protection risks; it is sensible to assess the feasibility of construction before expending resources on investigations for the design of the site. As much work as possible should be undertaken before breaching/retreating/tidal exchange, which may necessitate phasing works over more than one year. In assessing this, it is important to involve the (earth-moving) contractor to ensure that CDM considerations are included in the method statements and that construction is safe and practical. Any limitations imposed as a result of safe working requirements may necessitate modification of the design.

5.5.3   Creation of new shoreline

Realignment to rising ground

Where realignment is to naturally occurring rising ground, it will not normally be necessary to undertake work on the new shoreline except on environmental or amenity grounds. The effect of saline intrusion needs to be considered and existing land drainage may need to be effectively broken or parts of higher ground included in the scheme. Where the new shoreline is at the base of a cliff or the bottom of an escarpment, it will be necessary to evaluate stability by assessing the local wave climate in the area and the erodibility of the cliff or escarpment material. An increase in pore water pressures at the base of the cliff escarpment or steep slope could cause instability. Stabilisation measures, such as scour protection, may have to be provided.

Shingle ridge or dune destabilisation

Where the new shoreline is intended to be a sand dune or shingle ridge it may be appropriate to destabilise existing dunes or ridges (and/or the new intertidal area) to enable natural sediment movement as part of the scheme implementation. It will be important to have a clear understanding of the new position of such features if left to natural forces. If the site is not large enough to accommodate natural change fully then it may be necessary to allow for continuing management of sediment either in the short term (for example, prior to restabilisation by vegetation, or by using stabilisation techniques where a new ridge begins to develop) or to provide backstop measures to prevent impacts beyond the site. Such measures might include a bund to restrict the extent of further migration of a shingle ridge, or fencing to capture wind-blown sand at a suitable maximum realignment position within the site (allowing for dune growth at that location). Arguably, needing to control the extent of migration or growth of the ridge or dune indicates that the system is not self-limiting or sustaining, but it might provide a more sustainable environment than exists. Destabilisation of this nature has not been applied on a managed realignment scheme to date, so a precautionary approach should be adopted.

5.5.4   Treatment of new intertidal area

Changing land levels

In advance of the decommissioning of the existing defence and subsequent tidal inundation, land levels can be changed. Land levels can be raised or reduced locally by the redistribution of the material on site or by importing additional material to site, or exceptionally, by exporting from it.

When intending to lower levels it is important to be clear on the purpose of this and to account for natural sedimentation that may occur following implementation (see Boxes 2.1 and 3.2). To create mudflat, for example, may require land levels in the realigned area to be reduced. The material arising from this operation could be used directly to construct a new line of defence, to fill existing borrow areas in the site, to fill existing land drainage channels in the realigned area, or to construct a ring of higher-level habitat to landward. Lowering of land levels in this way will incur significant cost and may be avoided by selecting an alternative site for mudflat rather than by excavation. It should also be appreciated that if the land was previously reclaimed and is not low enough, natural processes might tend to accrete further fresh sediment on to the site.

The raising of levels can be achieved by using material excavated from, for example, borrow pits on the site, nearby freshwater ponds, or newly created creek networks that are part of the aim and purpose of the project. Warping of sediment using a tidal exchange system (see Section 2.7.3) can raise levels before inundation if there is available sediment in the system. The beneficial use of dredging sediments (from maintenance or capital work) might also be used to raise level either before and/or after realignment. Again, consideration should be given to the sustainability of any change made – if the placed material is immediately eroded from the new system it will have served little purpose and might cause negative impacts elsewhere.

Creek systems

Together with the decommissioning of land drains, the engineering of an initial artificial creek system within a realigned site (tying in with relict natural creek systems as observed from aerial photographic evidence if possible) can assist in the functioning of the system (French, 1995, 1996). Early establishment of a creek system enables tidal water and its suspended and bedload sediment to disperse into the realigned site and the tide to drain off the site. Creek systems also serve another drainage role, helping to dewater sediments. This may be important to allow sand sediment to dry out between tides to allow wind transportation for sand dune formation or to allow vegetation to become established in sediments that are not waterlogged. The initial creek system may change over time as a result of natural processes, and a network of minor creeks may develop. To initiate such natural change, small creeks can be dug with near-vertical sides that will collapse and feed sediment into the system over time.

PART III

Creek margins develop small levees that may contain significantly coarser sediment than that between the creeks. Levees form improved localised drainage and provide different environments for flora and fauna to establish. Monitoring at one managed realignment site has found that the banks and adjacent margins of new embryo creeks drain faster between tides and show an increase in stability and shear strength up to 30 times that of the surrounding deposited sediments (Watts, 2002). This difference in the physical environment has led to colonisation by samphire (*Salicornia*) on the edges of the embryo creeks, whereas it is not recorded on the adjacent sediments, which tend to be waterlogged. Waterlogged sediments can limit the lower extent of plants and the course of vegetation succession; waterlogging might be desirable if a mudflat or sandflat is required and undesirable where a saltmarsh or sand dune is desired.

Construction of a preliminary creek system, in the realignment area, may be especially desirable where the sediments are strong and resistant to erosion. In such sediments natural creek formation could take a very long time. Monitoring at one site has seen little evidence of creek incision over a period of six years, with creeks only developing once 20–30 cm of sediment has been deposited in the site (Garbutt *et al*, 2004). The nature of the original (agricultural) soil may help to explain this, as it can form an over-consolidated horizon within it. Regional differences in theses layers may be correlated to the quantities of detrital calcium carbonate in the soil. Calcium-deficient alluvium (for example in the Thames estuary region) appears to be particularly sensitive to the effects of a lowered saline water table (on reclamation), inducing the deflocculation of clays and the formation of a dense horizon, whereas carbonate-baring alluvium (for example in the Severn Estuary) is insensitive to pore-water salinity changes on reclamation, giving an absence of over-consolidated horizons (Crooks, 1999).

Figure 5.18   *New creek systems may need to be created to help the functioning of realigned area, particularly where sediments are consolidated and not readily eroded*

Design of the creek cross-sectional area and planform should also consider existing creek systems in the (local) natural environment for visual landscape reasons. Mirroring the natural environment will help blend the realignment into the landscape. In some instances the landscape considerations might encourage retaining the existing (linear) drainage ditches of a site, rather than excavating a new creek network. Archaeological guidance (Trow and Murphy, 2003) recommends not excavating relict creeks if they have traditionally been used (as moorings, for example), although it is likely that the physical processes will tend to exploit these features and erode them naturally over time. Where any creek system is being excavated a watching brief for archaeological interest should be maintained.

Existing vegetation

Dealing with the existing vegetation in the intertidal area will be site-specific and to a large extent dependent on whether the land levels in the area are artificially altered before breaching or bank removal occurs. There may be landscape reasons for retaining features such as trees and shrubs where they are at sufficient elevation, but experience has shown that these will die within a few months if their roots are flooded by saline water.

Figure 5.19   *Trees and hedgerows will die off when inundated with saltwater, so, unless there are good reasons to retain them, it is recommended that they be removed from the site before realignment*

For visual landscape reasons, therefore, it is advisable to remove trees and hedges within the floodplain before realignment. This will also reduce risks from dead wood floating into navigable areas. Hedgerows and trees can also disturb flight lines for waterfowl and may limit usage of the site by the suite of bird species that might be expected (Atkinson *et al*, 2001).

Other vegetation, particularly grasses, may be salt-tolerant and hence survive in the new environment. Experience has shown that grasses should be cropped very short to avoid mats of vegetation that will decompose and create an extremely oxygen-reducing environment. Cropping also encourages the trapping of seeds from the tidal waters. Leaving some vegetation also provides a surface roughness that helps sedimentary processes and hence vegetation colonisation. Experience suggests that establishment of saline-tolerant species can be retarded by more than a year if this pre-treatment (eg cropping or ploughing) is not applied (Reading *et al*, 2002). Where the intertidal area, in whole or in part, has been compacted during construction of the realignment scheme, the surface will usually require loosening by ploughing and/or harrowing. It would also be prudent to suspend the treatment of the intertidal area with agricultural chemicals for as long as possible before breaching or removing the existing tidal defence to prevent discharge of chemicals to the coastal environment following tidal inundation.

Figure 5.20   *Consolidated sediments may be loosened before realignment by ploughing and/or harrowing. This can provide a rough surface to encourage sedimentation and vegetation establishment*

In many instances, dunes and ridges, and the area landward of them, have been deliberately stabilised by vegetation or sediment management, or naturally stabilised by fine sediment filling interstitial spaces. Destabilising dunes might be possible by vegetation removal. Care will be needed not to cause damage to any desired habitat or other asset outside the site area, or to allow sediment mobility under initial conditions that will lead to sediment loss from the system. Care will also be needed to avoid the release of a plume of fine sediment or dead vegetation to the aquatic environment.

Buildings and structures

An area over which managed realignment of the tidal defence is to take place might typically be agricultural land devoid of any development. If there are any derelict or superfluous buildings or structures remaining, these would need to be demolished and the debris removed, or, if inert, buried to landscape part of the site. Inert material may be used creatively, such as in higher-level islands for bird roosting and nesting (these might also be formed of other local material) or in new footpath construction. It can be beneficial to public acceptance of a realignment to maintain (or improve) the recreation value of, and access to, the area, for example by providing bird hides, mooring areas, bike tracks, or establishing a new form of farming using the intertidal area. The ability to provide such access and enhancement, notwithstanding the project aim and purposes, might be limited by the need to avoid disturbance to local people or to wildlife.

Accommodating change

Over time the nature of the new intertidal area will change, for example the level may increase as a result of sedimentation and colonisation by vegetation will take place.

Therefore it is desirable at an early stage to develop a long-term plan for the intertidal area, which is sufficiently flexible to accommodate future change. It is also important to appreciate that some processes operate over long timescales, and a fully natural system may take decades or longer to develop.

Figure 5.21 *Managed realignment sites will develop and evolve over time. This site shows the situation three months after realignment and three years later*

5.5.5

Establishment of vegetation

In most situations, natural recolonisation is preferable to planting or seeding a managed realignment site. Natural colonisation will reflect the existing species and allow the range of species to change over time from initial colonisation to site maturity. This is likely to provide a range of plant species that can adapt to future change and are suited to the niche environment offered by the realignment site. Allowing nature to revegetate the site also reduces scheme costs in terms of investigating the ability of species to tolerate the site conditions, modifying the site to suit particular species, and planting or seeding costs.

There are some situations where natural regeneration will not occur, or not quickly enough, or cannot be allowed to occur, for example where:

- the site is remote from a natural seed source (especially seed sources for high-level saltmarsh plants)
- the necessary design of the site limits the amount of seeds getting on to the site (either attached to sediment or floating or wind-blown) – which might be the case for some tidal exchange systems, for example
- there is a danger of invasive or exotic species becoming established, making faster colonisation desirable (especially for sand dunes and gravel ridges)
- sediments will benefit from being stabilised by vegetation (such as sand dunes or where internally generated waves might prevent colonisation or on new flood banks)
- rapid colonisation is desired for visual landscape reasons
- colonisation by specific species is desired.

In these situations, seeding or planting may be necessary. Where possible, the source of seeds, plants, cores, turfs or sprigs should be from an existing marsh close to the realignment site to ensure genetic adaptation to local conditions. Many natural areas are protected by nature conservation designations, so permission will have to be obtained from the relevant conservation organisation before taking plant material or seeds from the sites. Table 5.4 provides a summary of guidance on approaches to seeding or planting saltmarsh and shingle ridge species (adapted from Brooke *et al*, 2000) that will help establish an effectively functioning system.

Table 5.4   *Saltmarsh and shingle (gravel) seeding and planting options*

Option	Commentary
Natural colonisation	Natural colonisation is the most desirable technique, and will usually occur over time without management. Seeds, sprigs, and rhizomes can float (on air or in water) and establish themselves on the new intertidal area. If there are no similar habitats or plant species in the immediate vicinity, colonisation is likely to take longer, as it will occur more gradually. Monitoring suggests pioneer species colonise within a year and achieve coverage in the second year after inundation (by breach or bank realignment).
Sprigs	Transplanting of sprigs of some species (such as *Salicornia* spp and *Suaeda maritima*) taken from a suitable existing (donor) site can prove successful given satisfactory physical conditions.
Turfs (saltmarsh)	Turfs, comprising rooted plant material and a seedbank, can be removed, stored and replanted firmly (or pegged down) at the same elevation as the surrounding area. Turfs are usually replanted when removed at the site (eg as part of engineering works). Turfs might be needed to establish vegetation on new earth flood banks quickly, prior to inundation.
Collection of strandline material	Driftline materials may be collected from nearby areas in autumn and transported for distribution at a (suitably protected) new site.
Cores (plugs)	Cores or plugs can be taken from suitable existing (donor) areas and transplanted to facilitate colonisation. These may also be grown on in artificial conditions (although this is expensive).
Vacuuming of seedheads	Seedheads at a nearby marsh may be "hoovered", stored (if necessary), and hand-sown.
Nurse crop	A nurse crop can be designed to provide cover to germinating and growing seeds. It can also be used where vegetation establishment is required to promote accretion and hence create conditions suitable for the establishment of other species.
Direct sowing of seeds	Direct sowing (drilling) is generally only suitable for high-level saltmarsh zone (and possibly well-protected mid-zone) species. There are some commercially available seeds suitable for this purpose.
Planting nursery-raised plants	This may need to be considered when a donor site cannot be used. Plants may be potted or bare rooted. Transplanting these plants should be done progressively, gradually increasing salinity if possible to avoid plant stress.
Seeded mats	Seeded mats (jute or coir) have been used in the USA, pegged on to the site. These mats could assist in trapping further sediment and seeds. They are not suitable in locations where wave or tidal action is likely to dislodge them. For flood banks this may also provide a suitable alternative. Non-seeded mats have also been used to reduce erosion of earth flood banks before vegetation naturally colonises.

Figure 5.22   *Natural colonisation of pioneer species one year after realignment. Note the beneficial affect of harrowing the surface before inundation*

If natural colonisation is not acceptable, the collection and distribution of driftline (strandline) material is probably the most cost-effective approach. Collection should take place in autumn, following seed generation, but before germination takes place in spring. It should be noted that thick rafts of strandline material decay and generate heat, causing germination to occur before winter and hence affecting survival rate. To avoid this, the driftline material should be collected as soon after seed production as possible. This material can be spread (broadcast) across the site. As the seeds remain dormant over winter, this may be feasible in advance of inundating the site in the following year. The best time for establishing seeds or plants is October to December and March to April inclusive for mid- and high-level species and springtime for low-zone (pioneer) species. This timing helps to explain why some managed realignment sites have not vegetated until the second year following inundation.

Table 5.5 provides a summary of guidance on approaches to seeding or planting sand dune species (adapted from Brooks and Agate, 2000). The planting (and management) of sand dunes is common practice and may be necessary to reduce mobility of bare sand if natural colonisation can not be allowed.

Table 5.5   *Sand dune seeding and planting options*

Option	Commentary
Dune grass planting (transplanting)	Fairly low costs, but labour-intensive and needs continuing management (Scottish Natural Heritage, 2000). Transfer existing established plants from one dune location to the desired area to plant; March is the best time to plant. It can be beneficial to use straw bales (or similar) in parallel lines, a few yards apart to enhance growth of vegetation and to protect it from immediate wind damage.
Planting shrubs	This may be used to protect windward sand faces in areas where the sand is too steep and mobile for grass to establish. Even if these shrubs fall over and die, they will still trap sand and contribute to dune development. Shrubs should only be used on sites where they already occur or where there is a need to retain open dune habitats. This is because they change the dune ecosystem more radically than grasses do and can invade grassed areas.
Sowing	Sowing can be used to establish a vegetation cover in areas where transplanting is impractical, eg where there is only a thin sand layer, or a large area has to be treated quickly. Sown seed may need some fertilising before it becomes fully established in a cycle of growth and decay; mulching, thatching or binding can help establishment. Sown seed takes about six months to germinate.
Fertilising	Fertiliser should be applied to unstable soils, as it helps to bind the sand together and reduce soil movement (using mulches). This increases the density of grasses at the expense of lower-growing herb and bryophytes. This may not be desirable for nature conservation, however. Fertiliser application should be kept to a minimum.
Thatching/mulching/ matting/binding	Thatching – covering of exposed faces with cuttings from scrub or forestry plantations – should prevent erosion, act as a sand trap and discourage trampling from people. Mulching – chopped straw/peat/leaf litter/reed/cut grass/seaweed/wood pulp/farmyard manure/sewage sludge helps bind the surface for grasses to establish. Matting – the material is made of straw filling sandwiched behind two layers of natural netting. It is pegged down to the sand surface and holes can then be made into this and grass transplanted on to it. The matting protects the surface and will decay in time. Binders – chemical glues have been used to prevent wind erosion of planted or seeded areas.

The Living with the Sea project (English Nature *et al*, 2003) provides details on saltmarsh vegetation succession in coastal and estuarine environments and also provides information on saltmarsh and sand dune vegetation management. The electronic guide (www[11]) has a detailed set of references to other documents for those who need to delve more deeply into the conservation management aspects of the realignment process.

Once the realignment is complete, subsequent development of the vegetation (and the associated species that colonise) may be greatly influenced by the nature of the management regime. The use of grazing animals and access to the site (resulting in physical trampling) is an important factor. These considerations should be identified as part of the approach to establishing and sustaining vegetation.

5.5.6 Design of breaches

One of the advantages of breach retreat is the fact that the existing defence remains, at least in part, as a structure that limits site exposure to wave action and so encourages sedimentation and vegetation before the outer wall ultimately fails. This has been the technique most commonly implemented in practice, despite frequent underestimation of increased velocities through the breaches, and hence of potential wider-scale impacts such as the widening or deepening of existing intertidal channels or the ultimate deposition of eroded material. It is believed that in some cases this relates to application of the English Nature formula to the breach width (see Section 5.3.2) and that better understanding of these issues will encourage adoption of wider breaches.

Figure 5.23   *Breach managed realignment has been the technique most commonly applied*

To ensure minimal adverse impact on the intertidal profile seaward of the existing tidal defence, the positioning of breaches ideally should coincide with the location of existing channels that cross the intertidal profile (or crossed it historically). These can usually be identified on the ground or from aerial photography; they also often coincide with the site of tidal sluices used for land drainage. These are a logical location for breaches, as the previous creek/drainage systems will provide a preferential route for flows both on to and off the site. If historic creeks are not present, or cannot be identified, then the existing land drainage system may produce a useful flow pattern. This may be desirable for archaeological reasons (showing previous land use), although it may provide an undesirably artificial landscape. Where multiple breaches are to be employed, careful consideration should be given to whether one or more breaches may be used preferentially (for example, because of existing drainage on the site, or topographic variation). Variability of the site may cause different locations and dimensions of breach to be determined.

Consideration should also be given to the number and positions of the breaches to control the distribution of tidal water over the managed realignment site. Two small breaches, for example, located at even spacing along the existing defence would give a more even distribution of tidal water than a single large breach situated at one end of the managed realignment. Where there is a large tidal prism on the site, multiple breaches may also be used to spread the impact on the ebb tide and where full bank realignment is not economically possible or practical to achieve (eg for CDM reasons). In designing breaches it is important to take into account the residence time of the water on the site and the degree of gravity slope that will be generated between the water level inside and outside the site. This not only has relevance to sedimentation and erosion, but also may be relevant to maintain the time period for land drainage to flow seaward (any reduction in time may increase flood risk and an increase in time might be beneficial). It may also be that other objectives can be met, for example, by providing more than one breach islands can be left between the breaches that may provide suitable nesting or roosting sites for birds.

Figure 5.24   *Two small breaches along the existing defence yields a more even distribution of tidal water than a single large breach. This can help in system functioning*

The bed level of a breach can vary significantly depending on the aims and objectives of the site. There may be instances where the level of the foreshore is higher than the new intertidal area. In this case it may be necessary to extend the breach across the foreshore so as to drain the new intertidal area on the ebb tide satisfactorily (unless a lagoon is desired). It may be desirable to leave a sill between the site and the natural environment, to enhance accretion or reduce jetting of water through a deeper channel. Over time, a deeper creek may incise itself and the consequences of this should be considered.

A smooth transition between the creeks/drainage within the new intertidal area and the foreshore and/or creek system in front of the existing tidal defence will minimise impacts to seaward and also mimic the natural environment in estuaries. Providing a route into the site in this way allows sediment transport into the site at lower phases of the flood tide, particularly for levee-building sediment. This allows the natural convex-concave profile between creeks to develop and contributes to the natural functioning of the system, and so is likely to lead to a more sustainable outcome.

Determining the width of a breach is an iterative process. An indication of the width that might be achieved under a "do nothing" approach, perhaps using the English Nature formula (see Section 5.3.2, Figure 5.3), will give a baseline condition for comparison. To use such a width in the design might constrain flows and cause scour locally so the design width can be increased to achieve the desired flow (and hence sediment transport) conditions. It may also be that breach width is determined by the aims of the project and the context, for example it may be desirable to have a wide and open breach on an open coast realignment particularly where dune processes are involved, or to remove a large section of defence for landscape reasons.

Flow through a breach can be readily calculated, using broad-crested weir principles, for a series of incremental progressions during a tidal cycle. This would typically use mean high water springs, but it could also explore the differences to mean high water neaps or even surge conditions dependent on the tidal range and sensitivity of the location. The rate and extent of inundation of the managed realignment site would be obtained from this analysis and the velocities through the breach would be used to assess the sediment transport (perhaps using bed shear stress to compare with the critical shear stress of the type of sediment existing at the breach site). If the rate and degree of inward sediment transport or erosion seaward is not appropriate for the aim and objectives, then the breach width (or additional/fewer breaches) will need to be considered and the process repeated. Once a breach width has been determined, a check should be made on the conditions prevailing during an extreme event to avoid negative consequences. It is also important that any local topographic variation (on the site or to seaward) is also accounted for that might cause flow concentration above the average conditions. The consequences of (non-armoured) breaches widening and/or deepening over time should also be assessed.

Figure 5.25   *Breaches that are not armoured may change over time. The breach design should account for such change as well as temporal variation of the flows at the site*

If the tidal prism *needs* to be constrained on the site (to avoid adverse impacts) and the tidal defence is composed of an earth embankment, it is likely that the breach will erode if it is not armoured. If this presents a problem, a tidal exchange system may be more appropriate to modify the site conditions and hence reduce the prism in a controlled way. On coastal sites this may not be an issue where a wider beach, open to the sea, is the desired outcome. Forming a single breach in a sand or gravel ridge, which will erode further, may be cost-effective and desirable to feed sediment into the system. This can allow the gradual erosion and movement of the existing ridge and will permit natural processes to reform any feature where it determines. This will be possible only where the site is large enough, is not limited by surrounding assets (on either side or to seaward) that need protection, and where the movement of sediment will be in the desired direction. Before adopting this approach, the short-term consequences on sediment movements, as well as the final outcome, must be understood.

If intervention is needed to avoid wave penetration, wave breaks may be located within the site in line with the breach(es). This approach can remove (or defer) the need for scour protection of any new realigned defence to landward and prevent re-erosion of sediments deposited within the site. These wave breaks may well be seen as sacrificial features that will gradually decay but in the meantime will allow the site to adjust to the tidal prism, accrete sediment and/or vegetate. Where the desired outcome is an open beach, waves should be allowed to penetrate into the site to redistribute non-cohesive sediments and form, for example, a storm beach or ridge in a new and more sustainable position on the intertidal profile.

5.5.7   Tidal exchange design

Tidal exchange systems are realignment techniques that allow tidal inundation to be controlled to a defined degree. They can be used in their own right or as precursors to bank or breach realignments in the future (Lamberth and Haycock, 2002).

Following construction, neither breach nor bank managed realignments normally include any control of the tidal inundation or egress from the site (although they have different effects on modifying this) whereas tidal exchange systems allow the (artificial) control of tidal ingress and limit hydrodynamic impacts in a definable way. The techniques are particularly suited to sites where:

- environmental impacts are uncertain
- saltmarsh vegetation is desired but the sites are currently too low in the tidal frame if exposed immediately to a natural flooding regime
- a habitat is desired but flood defences serve a function (within the wider system) that means they ought not be removed.

Once accretion has attained an acceptable level, it may be possible to dispense with the control systems. However, this may take some years to achieve, particularly where (suspended) sediment supply is limited.

Tidal exchange will involve the design of the control systems, usually comprising the lowering of the crest of the existing tidal defence to create a spillway or the installation of culverts or pipes through the defence. (If the wall remains as a flood defence, it should achieve appropriate design standards, as with any required new defence line inland.) The tidal flow is restricted by the sizing of these systems and further regulated by penstocks or sluice (flap) valves. Where tidal inundation of a site is by a spillway, a means of draining the site at low water may be required. This is likely to be in the form of a culvert or pipe controlled by a penstock, flap valve or similar arrangement. It is also possible to vary sediment loads entering the site by using control mechanisms set at different levels. Further control can be achieved by allowing water to flow in pipes set at various levels under and over each other to account for variable tidal ranges, locating the inlet and outlet pipes at different locations on the site, and using several pipes side by side that can be managed to provide a number of inundation levels.

The relatively small hydraulic capacity of spillways, culverts and pipes compared with defence removal or breach creation usually restricts their use to managed realignment sites of only few hectares in size, although they have been used in the Netherlands and Germany to control tidal flooding over extensive areas. The choice of materials for such construction should aim to reduce costs and avoid loadings where ground conditions may be poor (see Box 5.4). A further constraint is that the existing defence line has to be maintained for as long as the tidal exchange system is to function, so potential defence cost savings associated with breach or bank retreat would not be realised, or may be deferred until later. It may be possible to achieve some defence savings where the standard of defence can be reduced, for example where secondary defences are already in place.

PART III

Box 5.4   *Example of construction material selection issues*

In designing sluices or pipes, concrete construction has been avoided in some sites where it is a temporary measure to modify the landward area or where subsidence may be an issue. This can reduce the need for large foundations and the creation of a structure that is not easily removable. Polypropylene pipes have been used because they are durable, can be fused together, and will bend rather than crack if there is some ground settlement. Plastic pipe flaps have also been used to reduce costs but do require extra weighting to ensure an effective seal. Use of plastic flaps will reduce the need for steel and concrete infrastructure, so long as the scour at the pipe end is far from the remaining defence.

5.5.8   Land drainage considerations

The realignment of a tidal defence is likely to have some effect on the existing land drainage arrangements. The continuity of the land drainage system will need to be considered in the design. In many instances this may be achieved by the construction of a new drainage channel running along the landward side of any new realigned defence, but there may be wider implications depending on the way land drainage is designed and managed. The construction of a new channel would intercept existing watercourses in the area and discharge to an existing drain or through a new sluice in the realigned defence. This ensures the connectivity in the existing system both for land drainage and, where this is used by aquatic species, provides "assisted migration" of plants and wildlife into the new freshwater system. Care should be exercised in the location of land drainage sluices to ensure they do not silt up as a result of sedimentation in the realigned area or, conversely, cause undesirable scour. A new pumping station may have to be provided where the land level is lower to enable water to be expelled against tidal locking.

Continuity of land drainage will also need to be provided in the realigned area for the time between construction of a realigned defence and the decommissioning of the existing defence. Before decommissioning existing flood defences, any redundant culverts or land drains through, under or sub-surface to a realigned defence may require sealing, and the drainage channels within the realigned area filled, unless they are to be retained as tidal creeks or accounted for in the design. It is particularly important to consider any sub-surface land drains to ensure there are no connections to areas outside of the managed realignment site, especially where agricultural use or designated freshwater habitats may be involved.

5.5.9   Dealing with services

Where existing services cross the proposed realigned tidal defence and/or new inter-tidal area it may be necessary to divert the services around the area. Such services include electricity cables and telephone cables, water and sewerage and gas mains.

Where it is not practicable to divert a service, or where the cost would be prohibitive, then the effect of the scheme on the service will need to be considered and protection measures provided as necessary. Such effects include the potential for flotation, scour of the material above the service and additional physical loading on the service (for example from new defence works or inundation, or those that arise from natural accretion of the new intertidal area). Protection in the form of cellular concrete block mattresses or concrete rafts may counter the effects of scour and flotation, although the latter may also require some type of piling. The use of lightweight materials in the construction of the realigned tidal defence would reduce the increase in the loading on a service.

Overhead services retained across a new intertidal area will require access for future maintenance. It may be possible to provide such access at a high enough level for safe working. The bases of pylons designed for dry land may need to be protected if they are made from materials unable to withstand regular inundation (particularly from saltwater) or where the foundations might become unstable. It is essential to discuss potential impacts upon services with the owners/operators to determine appropriate courses of action.

5.5.10 Decommissioning of existing defence

Decommissioning of the existing defence should be built into the scheme costs. It is usually only possible to decommission once any new realigned defence has been constructed, has settled, become established and, possibly, become vegetated. In the case of earth embankments it would be usual for the realigned defence to be constructed in one earthworks season and the existing defence decommissioned in the following earthworks season, nine to 12 months later. This time delay would not normally apply when realignment is only to rising ground.

It may be possible to reuse materials from the existing defence in the construction of a new realigned defence, for example by reusing scour protection. This would need to be evaluated against the risk of leaving the existing defence vulnerable to wave attack before the realigned defence had become fully established. This risk could be minimised by transferring the material immediately before the existing defence was decommissioned.

To minimise working in intertidal conditions, the sequence and programming of decommissioning of the existing defence will need to be determined, especially in the case of bank realignment where the quantities of material are likely to be high. A staged approach is suggested, in which the existing defence is taken down in a series of layers, with the final, lower layers being pushed directly into an existing drainage channel or borrow pit situated immediately landward of the defence. In this process it is important to get final levels right, so as not to constrain the future evolution. It is important to determine the land level needed at the breach/old bank line to ensure that any required sill is at the desired level or that no sill is accidentally formed which may impair the design of the site. It is desirable that the final level is right first time, as re-entering the site once flooded may require more specialised (low ground impact) plant and machinery.

PART III

Figure 5.26   *Breaches can be designed to incorporate a sill to encourage sedimentation and/or reduce impacts to seawards. These may modify over time if not armoured*

It may be possible in breach realignments to leave or create an access route to the breaches along the existing defence line, but where there are more than two breaches in a site it might be difficult to access middle breaches safely (nevertheless, more than two breaches may be fundamental to the design). If more than two breaches are made, and any subsequent modification is required, it may be necessary to redesign and adjust only the outer edge of the breaches if access is limited. It is important to note that alterations to a scheme may need updated licences and permissions, if they have not been covered in the original design information.

5.5.11   Health and safety

Figure 5.27   *Construction of a realignment site needs careful planning to ensure safe working conditions for operatives and to protect the natural environment*

Throughout the development of a managed realignment scheme all activities should be carried out in accordance with current health and safety legislation and procedures. Apart from the "normal" health and safety issues during study, design and construction, the following are likely to need specific consideration and consultation:

- access to the remainder of the existing tidal defence following a breach managed realignment

- public exclusion measures in the new intertidal area and in the vicinity of each breach in the existing tidal defence

- fast-flowing water in the new/existing tidal creeks

- safe working for construction and/or maintenance taking account of tidal working

- temporary or permanent signage for the site.

Figure 5.28   *Signs may be needed to indicate alternative access routes and to alert the public to any dangers from realignment*

6 Monitoring

6.1 REASONS FOR MONITORING

Monitoring is an important aspect of any project, including managed realignment. It aims to measure a feature or range of features of the site to determine whether there is any change in that feature, to quantify or qualify the amount of change, and to inform coastal/estuarine management decisions. Monitoring a particular feature may require a range of parameters to be measured. For example, the geomorphology of a site may need topographic, sediment, hydrodynamic and vegetation parameters. Monitoring can feed into the design of managed realignment by providing data and information for use in assessment or to form part of an assessment where data and information is uncertain, for example to improve the accuracy of water level data and provide an assessment of return period water levels at the site. Pre-scheme monitoring, to provide a baseline condition, may also be required for an EIA and feed into an environmental action plan (see Section 2.5.1). The features that are assessed for an EIA may lead to an alteration of the design, to be able to mitigate or compensate for environmental change, or require assessment of change following implementation.

Monitoring has an important role at implementation and post-project in assessing the project's impacts and determining if the design is operating as intended. Implementation monitoring will usually be to observe and record features, such as archaeological finds, that become exposed during construction or impacts on, for example, shellfisheries. Post-project monitoring, and an associated action plan, may be set as a condition of a consent or licence (including within an EIA). The results of post-project monitoring might lead to a redesign (or some other form of intervention or even compensation) where an unacceptable outcome is shown to occur. Intervention should occur only where the degree of change is unacceptable (compared to a predefined criterion) and/or where an unacceptable change has been shown to exist for a long period. Sites will evolve over time, so the need to react to initial change should be carefully evaluated. Monitoring can also feed information about scheme design and performance into other projects.

Monitoring of the early managed realignment sites aimed to build up knowledge on their performance and to identify issues to feed back into future schemes. It covered a wide range of features and was intensive in nature. Adopting the same monitoring strategy may no longer be appropriate unless it is for research and development purposes. This section provides outline information on monitoring issues that are particularly relevant to managed realignment, which may play an important part in the overall scheme design and evaluation process.

6.1.1 Outline monitoring requirements

Monitoring may cover pre-implementation surveys, implementation and post project assessment. The initial surveys, often required for design purposes, may cover more issues, and be of greater detail, than monitoring to assess the performance of a scheme. It is important to identify the critical issues that *require* monitoring and to focus on *clear measures* that can help determine the success or failure of a scheme or that change it (see Section 1.6). Monitoring requirements will largely be dictated by the aim and purposes for the site (see Section 1.1) coupled with the requirements or conditions of consents and licences or mitigations defined in an environmental action plan (see

Section 2.7). The greater the risks identified, the greater the monitoring requirements are likely to be. The proportionality of assessment (see Section 3.1) needed to design the scheme also has a bearing when defining a monitoring programme.

A pre-scheme survey may be required as part of the data and information needed for site assessment and evaluation. It will also be necessary to begin monitoring of those parameters that will be evaluated post-scheme (see Section 3.2 for R&D projects that will provide further guidance on parameters and monitoring). It may also be important to monitor some parameters during construction, such as archaeological recording or visitor numbers. Post-scheme monitoring may also be needed to record, for example, items of archaeological interest exposed by any subsequent scour and erosion processes or ongoing visitor numbers.

It will be cost-effective to align pre-scheme design survey with the ongoing monitoring requirements. This might require a nested approach, where the detail of design survey is reduced to achieve the monitoring needs. In adopting this approach it is important to ensure consistency of data collection and, if no additional cost is involved, the greatest degree of data accuracy (rather than density or frequency) through the data set.

6.1.2 **Outline approaches to monitoring**

It is possible that monitoring for other activities will already be already taking place, so when defining a monitoring plan, synergy with existing activities should be sought. A monitoring plan should identify not only the parameters to be measured or evaluated but also the way that the data and information will be used. Any monitoring plan should be reviewed, probably annually, to ensure that features being measured remain relevant or that methods are modified where necessary. Modification might include the range of features/parameters monitored as well as the spatial extent and density/frequency of monitoring.

Without a control site, it may be very difficult to assess if the post-implementation changes have been caused by the managed realignment scheme. It can be hard to identify suitable control sites, because they may themselves be changing. It may be necessary to consider several control sites to reflect different features being monitored. If control sites are not used it may not be possible to put any negative impact into context, or further investigation/study may be needed should damage occur.

The recommended minimum information to collect on physical processes should enable the project team to determine whether erosion/accretion has increased or decreased both within the site and in the surrounding areas (within the system), and to assess the impact and consequences of hydrodynamic changes on morphology. Analysis can make use of biannual simple survey measurements (level data) and fixed-point photography. It will also be important to undertake a general visual inspection of the site defences (if any) to determine scheme performance and identify any remedial action that may be required.

Where physical data is limited, at least two years' pre-monitoring may be required for modelling and geomorphological assessment work. Some form of baseline must be established against which the post-monitoring can be judged. It may be possible to build a model (conceptual or numerical) and to test this to help identify the key weaknesses in information and data needs. The capture of new data can then be well targeted, can help refine the modelling approach and can be used in predicting the potential geographical area that a managed realignment scheme might affect.

If a certain type of habitat has been required as part of the scheme then monitoring the extent and type of plant colonisation will probably be required (see Section 3.2.1). Basic topography and repeat photography should be used as an absolute minimum, but quadrat or NVC surveying may be appropriate where this is a primary driver for the project. Study of vegetation should link to physical monitoring to help determine any limitations the sedimentary and hydrodynamic system has for different species.

Invertebrates have been monitored to demonstrate the overall health of the existing or new sediments under managed realignment and the potential food source for birds, fish etc. This has been applied particularly where SPA bird use needs to be monitored and there is concern over variability in bird counts as a result of other environmental factors beyond the realignment. Invertebrates, however, are expensive to monitor, so expert bird counts for the number and types of bird species might be employed for longer periods.

Landscape and historic environment monitoring can be used to demonstrate that certain landscapes are evolving or that information on historic infrastructure or archaeology is being recorded. Visual change should be recorded photographically and reported clearly where an impact may be developing, and *ad-hoc* monitoring may also be needed where a problem is perceived to be arising. Archaeological monitoring should include areas of the site that are beginning to erode (to determine if any interest is revealed). The requirements of this monitoring may align to other needs, such as for an AONB, national park or heritage coast, and should be integrated with such programmes. Requirements for such monitoring will usually be determined within the environmental impact assessment and may be a condition of a planning application.

If there are fisheries and shellfisheries nearby it may be necessary to monitor sedimentation in that location and/or suspended sediment locally. In one case, suspended sediment was recorded pre- and post-implementation to demonstrate the amount of change attributable to the managed realignment. Such factors may have to be monitored to determine if compensation for losses is attributable to the realignment.

It may be a condition of planning approval that numbers of visitors to the site are monitored. It has been suggested that managed realignment can both significantly increase and decrease visitor numbers. In one case, a major increase was noted immediately after implementation as a result of local publicity. Greater numbers might have consequences for local traffic, congestion and birds. Reduced numbers might affect local businesses. It may also be important to monitor effects to local channels for recreational or commercial navigation purposes.

It may be beneficial to monitor public perception of the scheme and to feed back information on progress to members of the public, particularly where the scheme has been contentious. Holding a public meeting can help assess perception (this may be quite subjective). Questionnaires could be used at these meetings to help quantify the level of public acceptance. For large or contentious sites, meetings to relay information and gauge perception might be held every six months for the first year(s), and then at project milestones or, following that, when new information is available.

6.2 PRINCIPAL MONITORING TECHNIQUES

Monitoring is an important tool when trying to assess the success of the scheme. It can also identify problems at an early stage before they become too large to rectify. Table 6.1 provides an explanation of the principal monitoring techniques applied to managed realignment schemes to date; further detail of specific tools and techniques are in Appendix 1.

PART III

Table 6.1 *Application of monitoring techniques*

Type of monitoring	Use of monitoring technique
Topographical survey	This is used to assess and monitor the ground elevation at or surrounding a managed realignment scheme. This can show, in general, which areas are accreting or eroding or stabilised. It also enables a judgement to be made on whether the conditions developing are suitable for a specific habitat etc. Topographical surveys can be carried out using land-based approaches or by aerial techniques, but survey accuracy needs consideration – eg because of sensitivity of saltmarsh vegetation to inundation.
Monitoring intertidal accretion rates	The importance of sedimentation/erosion of intertidal areas in controlling bed elevation of an area (and therefore vegetation coverage, as well as the overall morphological form) has meant that numerous workers have used a variety of techniques (Cahoon *et al*, 2000) to investigate these aspects in both natural environments and on several managed realignment schemes. The choice of technique depends on the accretion rates expected and the period over which measurements will be made. In natural environments, measurements have spanned timescales of single tides or weeks (Reed, 1988) to several centuries (Kearney and Ward, 1986).
Monitoring intertidal erodibility	The erodibility of sediments has not been systematically monitored at managed realignment sites. However, there is no reason to prevent such measurements being made and several approaches have been used to measure the erodibility of intertidal sediments in natural environments. Erodibility has been assessed by measuring a related geotechnical parameter, such as shear strength or by actually measuring the onset of erosion and increases in sediment in suspension. These measurements have encompassed both *in situ* tests and those where samples have been removed and tested within laboratories.
Flow monitoring	Variations in current velocities (speeds and directions) play an important role in determining sedimentation and erosion rates, both within managed realignment sites and in the wider environment. Flows also have importance for navigation and recreational use. It is often desirable to obtain a quantitative understanding of current velocities to assist in the post-project evaluation of managed realignment schemes.
Scour monitoring and counter-wall erosion	Regular surveys should be undertaken of the managed realignment site, the previously existing intertidal area in front of the site, the remaining part of the existing tidal defence, the realigned defence and appropriate areas upstream and downstream of the site. This information can be used to compare the actual development of a managed realignment site to that predicted at its time of implementation. Any differences identified should be investigated and taken into account when further managed realignments are planned.
Ecological monitoring	Depending on the particular objectives of the realignment, ecological monitoring is likely to focus on vegetation (diversity, abundance and percentage cover), benthic invertebrates (diversity and biomass) and/or birds (high and low water counts of feeding and roosting and possibly over-wintering birds). Ecological monitoring (see also Section 3.2.1) may be used to determine success where specific habitat or species requirements relate to a site and also to determine how a site contributes to the overall functioning of a natural ecosystem. Monitoring can also be useful in showing evolution of the site with changes in species types and abundance over time.

6.2.1 Outline programme for post-project monitoring

A post-project monitoring programme for a managed realignment site should be established and should include appropriate timescales for monitoring to be carried out. The frequency of monitoring should be proportional to the scale and nature of the scheme. An example of a generalised programme is shown in Table 6.2 to provide some indication of timing intervals. The specific features and parameters being monitored will need defining on a case-by-case basis for activities.

Table 6.2   *Example of a monitoring schedule*

Time after completion of managed realignment	Reason
Immediately after breaching (for first six months)	To capture quick and dynamic large- and small-scale changes occurring over the first few sets of tides and months.
Six months	To monitor large-scale changes occurring as a result of managed realignment implementation (eg formation of creeks, scour and erosion, sedimentation).
One to two years (depending on the size of scheme)	To check whether the site is "settling down" – ie changes are reducing and mirroring naturally expected levels. If not, there may be a problem that should be remedied to allow the managed realignment objectives to be met. It is important to monitor whether species have begun to colonise.
Years 3, 6 and 10	To monitor the physical development of the site, and to check that the habitat is developing as expected. Attention will begin to shift to the smaller-scale changes that are taking place (because large-scale change should not still be occurring).
After 10 years	To note gross change and development, but probably not including any detailed species monitoring. It may be advisable to return after 20 years to observe the changes over the previous 10 years. It is unlikely that a consent or licence will set a requirement beyond this timeframe, but at the 50- and 100-year points monitoring would contribute to understanding of long-term sustainability.

Box 6.1   *Example of post-project monitoring of vegetation*

The managed realignment site was completed in 2000, and post-project monitoring of the site was undertaken to determine whether it was successful in contributing to the status of a Special Protection Area (SPA). One of the recommendations of this project was that no more than 30 per cent of the site was to develop into saltmarsh.

The objective of the monitoring was to determine how the ecological value of the site was developing, and to what extent it was contributing to the SPA status. This objective was achieved by:

- recording the successional stages of the benthic community development. This included an analysis to determine the ability of the site to support specific bird species
- recording colonisation of the site by vegetation in terms of species abundance and distribution
- determining bird usage of the site.

The ecological development was monitored by measuring:

- sediment particle size distribution
- benthic macrofauna
- vegetation
- birds.

Because the main driver for the site was ecological, it was felt that vegetation monitoring would be sufficient to provide an indicator of level changes, so sediment accretion in the site was not measured directly.

PART III

MAXIMISING THE USE OF MONITORING DATA AND INFORMATION

In addition to meeting site-specific criteria, monitoring should be conducted with wider use of the data and information in mind, including use by the scientific community. A sound data and information management strategy should be employed (Millard and Sayers, 2000). Opportunities for working with other organisations should be investigated, because this will help keep costs down and perhaps provide the means to monitor aspects of the project beyond immediate requirements. This will promote collaboration and partnership and help develop broader scientific understanding.

A new Defra PAG (FCDPAG6) will address key features of data and information audit and management. This will provide information on the kinds of audit activities that might be appropriate to monitoring activity and help the wider use and sharing of data and information. Such activities might include checking:

- assumptions about data sets received and/or processed
- integrity of database(s) (including physical integrity of the storage medium)
- the size of monitoring database against previous expectations
- the degree of use of monitoring data based on access via the Internet or other forms of data supply.

It is also important to identify a programme for the collection, storage and dissemination of monitoring data. The programme might be influenced by:

- requirement of a consent, licence or agreement
- elapsed time, including promises made during the development of the scheme
- significant changes in the data set, which might modify the monitoring regime
- depletion of stocks of hard copies of data sets as a result of demand
- software obsolescence.

Monitoring of any natural environment carries risk and uncertainties and can be costly to undertake. A monitoring programme might have to be modified in timing, extent and intensity if it is to provide suitable information, and it should be kept under review and adapted or changed as necessary. When changing a programme, the effect on the usefulness of the historic information (for comparative purposes) should be evaluated.

It is important to deliver a professional output to the programme, particularly where the monitoring is part of a consent, licence or legal requirement. Trans-disciplinary groups are best placed to manage monitoring, using specialists in the relevant field(s) for each feature. This will ensure that the work carried out is scientifically sound. The data that are captured should have meta-data associated with them, and good practice for data and information management should be followed (see Millard and Sayers, 2000). The information generated from monitoring should be shared (see Sections 1.6.2 and 9.1 and made available to a wide range of stakeholders and researchers. This will increase knowledge about managed realignment and contribute to improved design for future schemes.

DESIGNING AND IMPLEMENTING MANAGED REALIGNMENT

1   Scheme design is an iterative process that involves the identification of a range of initial options or concepts, their assessment and subsequent refinement of design to reduce the impacts or risks. To undertake the design process successfully, specialist contributions should cover a range of disciplines, such as engineering, physical processes and morphology, natural environment, economics, and health and safety.

2   Various tools exist to assist in the assessment of physical processes associated with managed realignment schemes. These may be used at different stages of the design process to provide an appropriate degree of certainty in the outcome. Some of these tools are more applicable to estuarine environments, others to coastal environments. Often there will be a requirement to use a range of tools in combination to provide a hybrid approach.

3   Most managed realignment schemes will need a monitoring system, the nature of which will vary according to the drivers for the site, the geomorphological and hydrodynamic assessment, and the requirements of consents and licences. The greater the risks identified, the greater the monitoring requirements are likely to be. Monitoring programmes can include consideration of both physical and environmental parameters.

PART III

APPENDICES

A1   Monitoring techniques

Monitoring (see Chapter 6) is an important part of many projects and may serve a variety of needs. For example, it may be started in advance of implementation to provide survey data to aid design, it may be a requirement of a consent or licence including environmental impact assessment (see Section 2.7); it may be needed to evaluate the success or failure of a scheme (see Section 1.6); it might provide information to modify or redesign a scheme (if necessary) and it can provide useful information for the design of other schemes (where it is in a suitable form to be shared). This appendix is not intended as a comprehensive monitoring guide, instead it identifies some methods and issues that have been commonly used or encountered on managed realignment projects. The ability of a monitoring programme to meet its aims successfully hinges on the selection of an appropriate method, together with its deployment strategy, to measure each attribute effectively.

A1.1   TOPOGRAPHICAL SURVEY

Topography is important in assessing how a site may respond to physical and biological (ecological) processes following realignment. It might also be used for other purposes such as determining creek (or drainage) positioning, identifying the location of any archaeological interest, where it might be safe to re-route access or footpaths around a realignment site, or in assessing quantities for any new defences that might need to be built. Some of the main methods for measuring and monitoring topography are described below; all may be complemented by boat-based bathymetric surveys.

A1.1.1   Land-based methods

- *Total station* – a tripod-mounted system to sight and record the location in three dimensions of a staff-mounted prism at a remote, line-of-sight location. This can provide high accuracy of data – up to ± 1 mm dependent on distance surveyed.

- *Terrestrial laser scanner* – a tripod-mounted high-precision laser measurement system to survey the topography in three dimensions at a remote, line-of-sight location. Measurement resolution can be as high as 5 mm (depending on the range) with a measurement accuracy of around 50 mm.

- *dGPS* (differential global positioning system). This system determines its own position with respect to at least three GPS satellites and a local base station previously installed at a point of known location. The use of the base station improves the accuracy from between 10–30 m horizontally for non-differential GPS to 10–30 mm horizontally and 100 mm vertically.

A1.1.2   Aerial methods

LiDAR (light detection and ranging) – an aircraft-mounted laser system that measures the land level at a relatively dense spacing (typically at a 2 m horizontal grid coverage). The processed data has a vertical accuracy in the order of ± 150–250 mm. There are now also low-level flying techniques, similar to LiDAR, that provide greater accuracy (around ± 100–150 mm) but at higher cost. LiDAR has also been used for surveying bed conditions (through the water column) but may provide inaccurate results, especially in turbid conditions.

Photogrammetry – stereoscopic pairs of overlapping vertical aerial photos (typically with 60 per cent overlap). The process can be analytical (in which the images are viewed by an operator to identify land levels in a semi-automated process) or digital (in which automated stereo-matching algorithms are used to calculate land levels). The coverage and accuracy will depend on the photograph scale, scanning resolution and the quality of the block adjustment used to model the camera lens. Typically, vertical accuracies are similar and horizontal accuracy slightly worse than those associated with LiDAR.

SAR (synthetic aperture radar) – aircraft-mounted radar measurement system that measures the land level at a relatively dense spacing. The processed data has a quoted vertical accuracy of around 500 mm.

A1.1.3 Benefits and disbenefits of different topographic approaches

Carrying out topographic surveys by terrestrial means may be difficult and hazardous because of tidal conditions and difficult terrain seaward of a defence line, but they do provide high levels of accuracy in an environment where 100 mm can mean the difference between obtaining a saltmarsh or a mudflat, or where eustatic change may have a significant effect over time. For example, over a decade of monitoring 3 mm of eustatic change equates to 30 mm of level change. On a shallow sloping sand or mudflat this might account for, say, 3 m of horizontal change in that period. This would not be determined using less accurate monitoring methods.

LiDAR or photogrammetry may be preferred, particularly as part of scheduled strategic programme. This technique also provides a very useful visualisation of the site topography. National coverage of both digital photogrammetry and SAR topographic data is commercially available via the Web. A further advantage of photogrammetry is that it also provides a set of ortho-rectified aerial photographs containing a wealth of information about present-day habitat coverage, which can be compared with historic aerial photographs to establish changes in habitat distribution, morphological change, general sediment change and so on. Densely spaced topographic data (eg terrestrial laser scanning, LiDAR, photogrammetry, SAR) may also be used in conjunction with aerial photographs to produce 3D visualisations and animations of how the site may look following engineering works. However, LiDAR can, at certain times of the year, detect the top of crops/grasses rather than land level and therefore can lead to a significant underestimation of the actual tidal prism of a site, if not ground-truthed; such variation will not be linear across a site.

A1.2 MONITORING HYDRODYNAMICS

Water level is an important parameter to measure accurately at a proposed managed realignment site. The relationship between land (topography) and water levels is a key assessment in determining how a realignment might function. As with any parameter that varies over time, the longer the measurement record the better the assessment that can be made. However, it is feasible to monitor water levels for a short period of time and use these data to transform longer-term tide records when assessing water level. Apart from the water level, the flow (velocity) of water and the characteristics of waves are important in scheme design and, potentially, post-scheme monitoring.

Measurement of physical processes in deeper seawater is well established, but deploying and retrieving instrumentation can be problematic. Recording and transforming data from deeper water to the shoreline is complex and may misrepresent the actual conditions at the shoreline. To overcome this it may be necessary to deploy instruments in shallow water to provide direct data for input to assessments and baseline data for monitoring. Deploying many instruments in shallow waters, over intertidal flats where water depths may well be less than 1 m, presents an even greater challenge. Many of these areas are remote and difficult to reach with vehicles (except hovercraft) or on foot. However, techniques have been developed to provide data that may require greater or lesser technical skills in their application. Advice should be sought from experienced individuals to assess the risks, accuracies and practical requirements for undertaking such work and in analysing and interpreting the results.

The instruments described below are typically deployed for 15–30 days, to gain an understanding of the currents experienced over a spring-neap tidal cycle. Deployment periods can be shorter or longer, depending on the data collection method. If data are stored on a logger for downloading at the end of a deployment, for example, the length of the deployment is directly reliant upon the memory of the logger. Another method of data collection is through the direct relay of the data to a computerised database.

A1.2.1   Monitoring flow

Instrument types for measuring current velocities include:

- float and drift recorders
- direct-reading instruments that "tune-in" to the current, such as impeller meters
- instruments that measure velocities in one, two or three perpendicular axes, such as electromagnetic current meters.

Background information on available techniques follow below, but more details on flow measurements can be found in Dyer (1979) and Ingham and Abbott (1992).

Dye tracer

It is possible to monitor flow patterns in the short term using dyes injected into the environment. These provide a way of determining the direction and degree of mixing in the flow as the dye disperses from the injection point. It may be possible to use different dyes to provide a time sequence of flows, but this technique is quite limited in its temporal application.

Drogue tracking

The simplest form of flow tracking is to float an orange or apple on the surface and watch its movement; drogue tracking is a more sophisticated version of the same method. It is a relatively easy and cost-effective way of obtaining information on the pattern of surface, and near-surface, circulation systems over an area of water. Drogues should be designed to present minimum wind resistance and to represent water movement at a selected distance below the water surface. They can be placed in the water and their movements monitored regularly from land-based survey points. Surface drogues have typically been used for oceanographic studies (Hansen and Poulain, 1996), but, given sufficient depth, they could also be used to determine general patterns and rates of flow through a breach.

Impeller meters

The rotation of the impeller is proportional to the current speed, with the exact relationship being determined by laboratory calibration. The impeller meter can be mounted on a frame that is fixed on the bed or suspended on a cable from a boat or observation station on the water surface. In the latter case, vanes on the rear of the meter body align the instrument to the current, the direction of which is then measured with a compass.

This instrument is capable of measuring current at a set point through time. Care must be taken that seaweed or other obstructions do not foul the impeller, since the measurements are dependent upon its free rotation. An additional disadvantage is that vertical water motions can cause the impeller to turn, resulting in the inaccurate calculation of the horizontal current. Wood *et al* (1998) describe the deployment of bottom-mounted meters in the Humber Estuary. In this case, 10 impeller instruments were mounted on a pole along the flood-ebb axis to measure the vertical profile of currents for the flood and ebb tides.

Electromagnetic current meters

Electromagnetic current meters (EMCMs) operate on Faraday's law of electromagnetic induction. The physical arrangement of the electrodes used may vary between coils through which flow moves or electrodes about which flow takes place located on the surface of a solid discus with energised coils (for 2D measurements) or a solid sphere with energised coils (for 3D measurements). The liquid serves as the conductor and energised coils outside a flow tube create the magnetic field. The voltage produced is directly proportional to the current magnitude. Two electrodes mounted in the pipe wall detect the voltage, which is measured by a secondary element. The major advantage of these instruments is that they can measure forward as well as reverse current with equal accuracy (approximately ±1 cm/s). Additionally, as EMCM can record the *x* and *y* components of velocity, a measure of turbulence can be obtained. The main disadvantage is that these instruments tend to be expensive and, like the impeller meter, record currents at a fixed point.

EMCMs have been extensively deployed in various intertidal studies. Examples of their application in shallow mudflat environments include: (i) Christie and Dyer (1998), who deployed four instruments to determine the current velocity throughout the water column at four locations across a mudflat; and (ii) van Proosdij *et al* (2000) and Whitehouse and Mitchener (1998). In both the latter studies several instruments were attached to a bed-mounted frame to determine the velocity field at one location.

Acoustic current meters

The most common type of this meter is the acoustic Doppler current profiler (ADCP). This versatile, high-precision instrument can measure two to three current velocity components. The current speed and direction are measured by the emission of acoustic pulses. Analysis of the returning frequency of these pulses allows the velocity to be calculated. The measurements are insensitive to water quality, which allows for a wide range of applications. The device's major advantage is its ability to measure currents at a range of depths simultaneously. The instrument can be deployed in a fixed (Eularian) mode or a moving (Lagrangian) mode. It may be mounted to be downward-looking, measuring from the water surface to the seabed, or upward-looking, from the bed to the surface. Interpretation of the backscatter may also allow suspend sediment concentrations to be determined.

Use of ADCP may be limited in shallow-water sites, dependent on the head array used and depth of water. They are useful in coastal/estuarine main channels to monitor flow and sediment profiles and provide important background data. ADCP devices were used in the recent the EA-funded STONE (Sediment Transport across the Ouse/ Newhaven Estuary) project in water less than 6 m deep. Another example of ADCP application can be found in Schettini (2002).

Rotor-type meters (Aanderaa)

These are made up of a series of paddles, each of which comprises two half-set half cylinders mounted upon a vertical axis. Small magnets on the ends of the rotors induce pulses of currents that are measured and these are proportional to the current speed. Although this instrument is more suited to use in deeper water environments (greater than approximately 10 m water depth), it is capable of measuring currents as low as 2 cm/s. However, this instrument is unable to distinguish direction when the current alternates rapidly.

A1.2.2 Monitoring waves

The monitoring of waves can be important in the design and implementation of a managed realignment scheme for two reasons: (i) to provide input design parameters (eg for purposes of assessing issues such as overtopping discharges of any required flood defences); and (ii) to enable assessments to be made of the degree of exposure of the managed realignment site to wave action, and hence the potential for re-suspension of deposited sediments.

Wave monitoring techniques can be grouped into traditional *in-situ* instrumentation approaches and the more recent remote techniques such as satellite altimetry. The latter are better suited to ocean-scale studies and still require calibration against measurements *in situ*. The following section therefore focuses on *in-situ* approaches.

Offshore waves

To characterise the wave regime of the general marine environment in which the scheme is set, offshore waves could be monitored by use of surface-following buoys, such as the Waverider buoy. These devices can be either directional (recording wave height, period and direction) or non-directional (wave height and period only) and typically are deployed in deep water. As offshore waves propagate towards the shoreline they are modified through processes such as shoaling breaking refraction and diffraction, so that the nearshore wave environment may differ substantially. Consequently, while offshore wave conditions provide useful information on general exposure conditions (and often are essential input data to numerical modelling), it is the nearshore conditions that become more relevant to the design of managed realignment schemes.

Nearshore and inter-tidal waves

Typically, waves in the nearshore and intertidal environment are measured by means of either pressure transducers mounted on bed-frames or by surface-piercing wave staffs. The latter should be arranged in arrays so that directional assessments can be obtained (Barber, 1963; Haubrich, 1968; Chadwick *et al*, 1995a and b). Simple measures of wave magnitude might also be achieved through locating fixed survey staffs in the intertidal and videoing them, the data then being generated by visual observation of the staff. Such approaches are useful means of determining nearshore wave parameters such as wave height, period, direction and the total energy spectrum. Wave parameters are important in the design process, but they may require monitoring to determine any changes as a result of the scheme. Changes might include alteration of nearshore bathymetry and, when arranged in a suitable configuration, assessments can be undertaken of the degree of wave attenuation across the intertidal profile. This may be of interest in the monitoring of managed realignment schemes, for example to investigate the attenuation properties of newly created saltmarsh or to determine if wave energy has increased on the open coast adjacent to a realignment site.

A1.3   MONITORING INTERTIDAL ACCRETION AND EROSION

A1.3.1   Monitoring intertidal accretion

Accretion is the *net* gain of elevation of a surface as a result of sediment deposition. It should be recognised that the *actual* amount of sedimentation occurring, being eroded and redeposited might be much greater than this change and have important mplications for ecological development and thus the use of a site by different species. Changes to non-vegetated surfaces are likely to be more dynamic than those where vegetation is established, but it must be recognised that levels can change even where a surface is vegetated.

The rate of accretion is an important indication of the evolution of a site (whether open coast or estuarine) and has implications for the sustainability of habitat it provides. Accretion of sediments may indicate that an effective beach is developing and responding to different events during an annual cycle or to longer-term changes. Accretion rates can be compared to changes such as sea level rise predictions to indicate if a site can adapt to sea level rise in the longer term.

Artificial marker horizons

Artificial marker horizons are commonly used to measure accretion rates in the 0.5–1.0-year timescale. Materials that have been used as marker layers for measuring accretion include natural-coloured sand (Oliver, 1929; Stoddart *et al*, 1989), dyed sand (Nielsen, 1935), brick dust (Stearns and MacCreary, 1957) iron filings (Chapman and Ronaldson, 1958), glitter (Harrison and Bloom 1977; Stumpf, 1983), glass-fibre resin and sand (Letzsch and Frey, 1980), and the disruption of silica plugs (Inglis and Kestner, 1958). Accretion is usually measured after a known time upwards from the marker layer (Ranwell, 1964). The depth of these horizons is measured by taking cores, digging trial pits through them or, in the case of rigid layers, by inserting a probe into the sediment. The longevity of such horizons is limited by their lateral extent combined with the need for destructive sampling and (for soft marker horizons) biotic sediment disturbance. Although the use of some sort of marker horizon provides an accurate measure of accretion, it is open to the objection that the surface microfauna are disturbed (Ranwell, 1964).

Other workers have measured accretion/erosion using the burial/exposure of aluminium plates (Collins *et al*, 1981). Using expanded aluminium plates allows normal processes to continue (such as biological disturbance or infiltration) while providing a horizon that will last some years. A small pit is dug and the plate inserted horizontally into the undisturbed sediment or it is placed in the bottom of a pit and the sod replaced. Normally a six-month settling-in period is allowed before measurement is made using a thin probe, replicated at least 10 times.

Sedimentation-erosion tables

In the USA, Boumans and Day (1993) developed a sedimentation-erosion table (SET) (Cahoon *et al*, 2000) based on earlier work by Schoot and de Jong (1982). These devices consist of a fixed foundation that is embedded at the site of measurement and a detachable horizontal arm that fits into the foundation and allows the elevation of the sediment surface to be measured from a fixed height. The elevation measurements are usually made by means of several metal pins, which slide down through the table arm to reach the sediment. Sedimentation-erosion tables are a good way to obtain accurate and precise elevation change at a limited number of locations. By comparing measurements of elevation change with separate determination of accretion using markers, additional insights into processes can be obtained (eg sedimentation versus compaction).

Sediment traps

Sediment traps can also be used, particularly for intensive short-term process studies, but not for longer-term monitoring programmes. Over individual tides, pre-weighed filter papers can be fixed to a mounting plate, which then is attached to the sediment surface. These papers can be collected after a given time, dried (to remove water) and reweighed to determine the mass of sediment deposited. If a density is known, the mass can be converted into a volume, from which one can calculate the thickness of sediment deposited over a known area. However, such an approach seals off the sediment surface from the stabilising microfanua within the sediment, such as diatoms, and this is one reason why measured accretion rates may be lower than actual rates (French *et al*, 1995).

Tracer sediments

In sand and gravel environments it may be possible to use tracer sediments to determine the direction and degree of sediment movements. This technique can range from sand dyed with fluorescent material, which is recorded using ultra-violet light, to smart pebbles that mimic the density and shape of gravel and emit a trackable radio frequency. It may also be possible to use coarser sediments imprinted with a magnetic signature, but it may be difficult to differentiate them from the background sediments. These techniques help to show how coarser sediments are behaving and can be useful in monitoring the way a beach is evolving compared with the predictions.

Topography

Topographic surveying can also be used to record intertidal accretion rates, although it can be difficult to measure precisely the same x–y location each time. To reduce these errors, transects can be set up and resurveyed at intervals, although slight offsets from the transect may produce errors that exceed the variable being measured (Carr and Blackley, 1985). The accuracy of topographical surveying compared with the anticipated accretion rates (order of millimetres or centimetres per year) means that this technique works best in areas of high accretion over longer time periods (eg several years). Repeat survey transects do, however, have the benefit of showing the development of the morphological form along a profile line or over a grid and can be coupled with aerial images to aid interpretation of site evolution. A simpler approach, which can be combined with topographic surveying, is to measure the position of the sediment surface from the top of the stake using a tape measure (Ranwell, 1964). The size of the stake and its depth of burial often determines how long the stake remains, although human interference can be a problem in populated areas and scour usually occurs around the stake. Another approach is to take a large number of random points across the realignment site and to consider the elevations in a statistical fashion, calculating average rates of intertidal accretion.

Datable horizons

Over longer time periods, sedimentation rates can be obtained if one is able to date specific horizons within a sediment core. This layer may be used as a form of datum, above which the amount of change can be determined. Radio-isotopic dating of sediment cores, for example using ^{210}Pb, can be used to obtain long-term rates of sedimentation over periods of 50 to several hundred years (Stevenson *et al*, 1985). Another approach is to measure the profile of contaminants within sediment cores, to date certain depths within the core. Allen and Rae (1988) employed this approach in the Severn Estuary and elsewhere, focusing on metal pollutants using the measurement of ^{137}Cs spike, which marks the onset of atmospheric atomic weapon testing in 1945 (Carr and Blackley, 1986; Bricker-Urso *et al*, 1989; Bryant and Chabreck, 1998; Anisfeld *et al*, 1999; Brown *et al*, 1999). Such techniques may be useful to understand both the historic and future evolution of a site.

A1.3.2 Monitoring intertidal erodibility

Testing the erodibility of sediments can help determine their mobility under physical processes, and the degree to which new sediments are stabilising in a newly realigned area. Erodibility can be measured by removing "undisturbed" sediments and testing the unconfined shear strength of the samples in laboratory-based equipment (Pestrong, 1969). Mudflat samples from the field have also been tested in laboratory flume devices to measure their erodibility (Amos and Mosher, 1985). The *in-situ* shear strength can be measured quickly and easily in the field using shear vanes. More recently, technological advances have allowed the development of mini erosion meters (Tolhurst *et al*, 1999; Tolhurst *et al*, 2000). Larger-scale field measurements have been made using frames that hold variety of equipment, including velocity meters and suspended sediment meters such as optical backscatter devices, to record the onset of erosion under flows *in situ* (eg DeVries, 1992). Such equipment arrays can be deployed from vessels or assembled at low water.

A1.4 GROUND AND FLOOD DEFENCE INVESTIGATIONS

It is important to understand any existing ground conditions from investigation reports and information about the history of the construction and maintenance of any structures. General information about the variation of the soils around the study area, including indications of the extent of former river/estuarine channels and floodplains is also required to help evaluate how a site might evolve once realigned and to monitor how such variations effect scheme performance. A national database showing the available ground investigation data is held on the BGS website (www[16]). Local geological societies may also have records that can be consulted.

An inspection of the site should include a structured walkover survey of the existing defences and identification of potential areas where the future defence (if needed) could be located. This will help in understanding, and monitoring, how defences decay over time and allow such change to be related to the evolution of the site.

In the course of the survey, the following should be noted:

- long profile of the embankments – variations in level, fissures
- cross-section of the embankment – undulations, slips
- animal holes, indicative of potential structural undermining
- types of vegetation on the rear face and on the landward side, indicative of moisture contents and saline intrusion.

Weak points in a defence may indicate historic creek positions or locations that will preferentially fail, and this will help identify potential locations for breaching or failure in the future. A first assessment of embankment stability can be made from standard charts. This will require certain assumptions regarding the existing ground conditions, both underlying and within the embankment.

A1.4.1 Ground investigation analytical techniques

More detailed analytical techniques will require design parameters to be calculated based on ground investigation data. The two principal considerations are embankment stability (the likelihood of collapse) and settlement (the degree of past/future lowering of the crest due to consolidation of the embankment and underlying strata).

Typical techniques that might be employed, in increasing order of complexity, and therefore price, are:

- window sampling
- hand-augering
- penetration shear vane testing
- piezocone soundings
- soft ground boreholes and piston sampling.

A1.4.2 Scour and erosion monitoring

The following methods can be used to assess the scour or erosion of/around defences:

- topographic survey
- hydrographic survey
- aerial (and other photographic) imagery.

Hydrographic survey would usually be undertaken in areas that are permanently covered by water and comprises the determination of bed profiles using echo-sounding techniques. Aerial imagery would include photographs or LiDAR that are taken regularly (say annually), giving a visual record of the manner in which the managed realignment is changing with time. Stereoscopic pairs of photographs can provide general topographic survey information and allow features such as creeks, pans and vegetation to be mapped.

Initially, a baseline would be established immediately before breaching or removal of the existing tidal defence. Surveys at the same locations and using the same methods would then be carried out at regular intervals. Continuous comparison between the surveys would provide information on:

- the development of the tidal creek system in the realignment area
- the development of the tidal creek system in the previously existing intertidal area in front of the site
- scour or sedimentation at fishery sites
- scour or deposition in navigation channels
- settlement of the realigned defence
- the rate and the manner in which the remaining parts of the original defence deteriorate.

A1.5 ECOLOGICAL MONITORING

Ecological monitoring can be a requirement of an environmental impact assessment, consent or licence, but it also shows how the environment is evolving and how sustainable habitats might develop over time. Monitoring information from local realignment sites might help identify the types of habitat that will form in a realigned area. Ecological monitoring has been undertaken at several experimental intertidal habitat creation sites, including Tollesbury, Essex (English Nature, Environment Agency and Defra), Saltram, Devon (Reading *et al*, 2002), Orplands in Essex (Environment Agency and HR Wallingford) and Northey Island in Essex (National Trust, English Nature and Environment Agency). The Environment Agency typically lets five-year contracts for monitoring at intertidal habitat creation sites, with data gathered twice a year during the contract (ABP Research & Consultancy, 1998).

Ecological monitoring need not comprise a long list of parameters, but it should include measures that reflect system structure and function, such as species diversity and community composition. This might be undertaken to NVC (National Vegetation Classification) standards for long-term monitoring, using the random quadrat approach. More detailed change can be identified using fixed quadrats. As a minimum, fixed aspect photography of the site will be needed to show development of species and allow some determination of the vegetation composition; this might be coupled with use of aerial photography to map the extent of developing vegetation. Related physical parameters are also important to understand the evolution of the site, specifically elevation (to measure rates of accretion or erosion) and sediment (type, grain size, pH, salinity, temperature, organic content and contaminants).

In some cases, fish monitoring may also be appropriate, but to gather meaningful data it would need to include the role of the habitat in providing food, shelter or spawning areas rather than simply the presence of fish at high tide. This can prove useful demonstrating that realignment can benefit fisheries. Some managed realignment projects have also had to undertake detailed work on suspended sediments, pollutants and sedimentation to demonstrate the impacts on local shellfisheries.

Project FD1918 – Habitat quality measures and monitoring

FD1918 is a Defra/Environment Agency, Flood and Coastal Defence R&D project on the provision of guidance for the monitoring of managed realignment where habitat creation is a key driver (see Section 1.2). Such sites cover the intertidal regions of both estuaries and coastal zones and include saltmarsh and mudflat habitat. The project will provide guidance for:

- the collection of better data in terms of ensuring its relevance, consistency and statistical validity
- the assessment of the success of habitat creation schemes
- validation of the effectiveness of mitigation schemes and assessment of the residual impacts.

Additionally, the project will help managers take corrective action where habitat quality objectives have not been achieved, or to develop alternative quality objectives that better reflect the capacity/capability of the site.

Habitat creation schemes are undertaken for a number of reasons, so the parameters to be monitored vary between schemes. The main reason may be to mitigate a particular loss to a designated site, or to form a general habitat type to compensate for loss from sea level rise. What is monitored and the level of detail required depends largely upon the scheme's overall purpose and the requirements for attaining a set standard of habitat(s) and/or support associated species.

A review of the factors important to habitat structure and function should be completed for each site. This emphasises the need to consider the physical, chemical and biological characteristics of a site along with the linkages that exist between each of these parameter types. It is also important to consider the functioning of the ecosystem overall (the interrelationships and symbiosis), not just the individual parameters.

A1.6 SUMMARY OF ENVIRONMENTAL MONITORING TECHNIQUES

By reviewing the case studies, it is possible to identify the techniques that have been used to monitor managed realignment sites (whether implemented or not). Existing methods that have been applied to situations capable of measuring the relevant parameters have also been identified. A summary is presented below.

Table A1.1 *Summary of parameters and possible monitoring techniques that could be used at managed realignment sites*

Category	Parameter	Monitoring technique
Morphology	Elevation	Differential GPS
		RTK GPS
		Established benchmarks
		LiDAR
	Topography	Differential GPS
		EDM
		Descriptive profiling
	Area	Cartographic exercise
		GIS
	Creek systems	Aerial photography
		Site mapping
Hydrodynamic	Tidal range	Tide gauge
		Photographic records
	Current velocities	Current meters
		Surface drogues
	Wave action	Wave recorders, wave staffs
Water quality	Salinity	Salinity probe
	Dissolved oxygen	DO probe
	Turbidity	Turbidity meter
		Water samples collected and analysed
		Secchi disc
	Contaminants	Collection of water samples
Sediments	Sedimentation rates	Sediment accretion plates
		Marker horizons
		Filter papers
		Sedimentation erosion tables
	Contaminants	Sediment cores
	Salinity	Refractometer or conductivity
	Water content	Theta probe
		Suction pressure measurement
	pH	pH meter
	N	Sediment cores
	Organic matter	Sediment cores
	Redox potential	Redox potential tester
	Surface cohesive strength	Shear vane
		Cohesive strength meter
	Particle size	Particle size analysis
Ecological	Vegetation	Aerial multispectral remote sensing including CASI
		Aerial photographs
		Measurements within quadrats
		Photographic records
		NVC mapping
	Benthos	Sediment cores
		Grab samples
		Epibenthic trawling
	Invertebrates	Suction sampling and pitfall traps
	Birds	Visual observation and recording
	Fish	Netting
		Traps
	Freshwater	Pond dipping
	Mammals	Count burrows
		Count latrines and droppings
	(and lizards)	Trapping

A2 Understanding specific landform processes and habitat (re-)creation

Understanding of different landforms is an important aspect in designing a managed realignment project. This appendix is not intended as a comprehensive guide to all the landforms and their associated processes and habitats. Instead it identifies some of the key environments and provides outline information on issues that have already been considered and used in implementing managed realignment sites. This means that there is an emphasis towards saltmarshes, as few other landforms/habitats have actually been created or re-created through realignment.

To address sites that intend to form something other than saltmarsh, implementation needs to be followed by development of an understanding of the issues faced. It is important to approach such sites through a thorough understanding and detailed researching of these types of environment. Only a cursory description of these environments has been made here.

A2.1 SALTMARSHES

Box A2.1 *Key references for saltmarshes*

> Adam, P (1990). *Saltmarsh ecology*. Cambridge University Press, Cambridge
>
> Allen, J R L and Pye, K (1992). *Saltmarshes: morphodynamics, conservation and engineering significance*. Cambridge University Press, Cambridge
>
> Boorman, L A (2003). *Saltmarsh review. An overview of coastal saltmarshes, their dynamic and sensitivity characteristics for conservation and management*. Report 334, JNCC, Peterborough
>
> Brooke, J, Landin, M, Meakins, N and Adnitt, C (2000). *The restoration of vegetation on saltmarshes*. R&D Technical Report W208, Environment Agency, Bristol
>
> Carpenter, K and Brampton, A (1996). *Maintenance and enhancement of saltmarshes*. R&D Note 473, Environment Agency, Bristol
>
> English Nature, Environment Agency, Defra, LIFE and NERC (2003). *Living with the sea. Coastal habitat creation: towards good practice* [www[11]]. English Nature, Peterborough

Saltmarshes can form in a variety of environments but require wave/tide energy levels to be reduced sufficiently for finer sediment (mud and sand) to be deposited and remain relatively stable. Saltmarshes may form on the margins of estuaries where the tidal and wave forces of the main channel are sufficiently reduced, or behind some protective landform on the open coast such as a wide intertidal flat, spit, barrier beach, shingle ridge or sand dune. Once the physical conditions allow vegetation to become established, the vegetation itself then helps further stabilise the environment and alter the landform as it evolves.

Saltmarsh has been the most common habitat to try to (re-)create through managed realignment. It provides ideal conditions for many species and so is often protected through national and international legislation and designations. To achieve a realignment design that will create conditions suitable for saltmarsh a variety of factors need to be taken into consideration. The following information aims to highlight common issues identified through reviewing managed realignment schemes.

Figure A2.1   *Saltmarsh habitat*

A2.1.1   Saltmarsh elevation and tidal inundation

The frequency of inundation is an important factor in determining the extent and type of saltmarsh that may form. Inundation is essentially a function of "land" elevation with respect to the tidal frame (range of tidal elevation and frequency and duration of inundation), and hence elevation is the most important factor determining suitability for saltmarsh creation. Tidal range and saltmarsh ecology vary between different estuaries and coasts, so the most reliable way of determining the preferred elevation range for saltmarsh at any given location is to survey the distribution with respect to elevation of existing saltmarsh at nearby locations (Burd, 1995).

It is possible to develop a model of the likely saltmarsh area using recent colour aerial photographs to map the distribution of existing saltmarsh and topographic survey data to record corresponding elevations. By entering both sets of data into a geographic information system, a histogram of saltmarsh distribution with respect to elevation can be obtained and applied to the site (see Figure A2.2). Outlying data should be discounted, as a small proportion of both level and saltmarsh distribution records are likely to be the result of mapping or interpretation errors. It is recommended that the 5th and 95th percentiles should be taken as representative of the preferred altitudinal range of saltmarsh at a site. This technique can also be used subsequently to present, and compare, the results of monitoring.

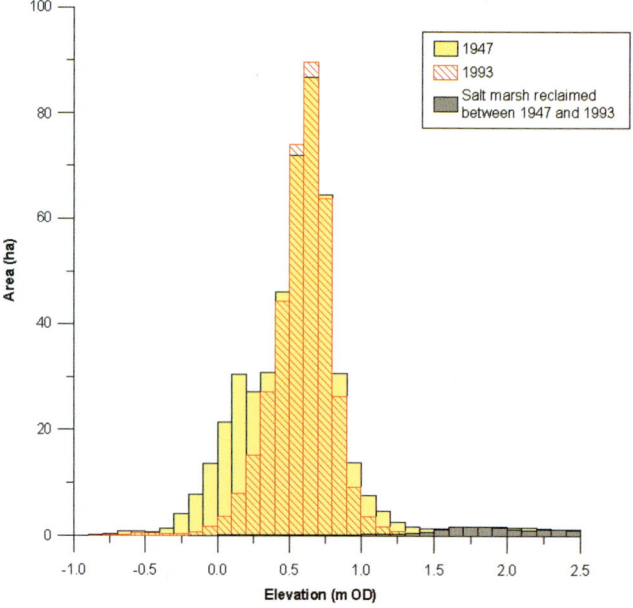

Figure A2.2   *Distribution of saltmarsh with respect to elevation in two different years*

It is possible to determine likely saltmarsh extent using saltmarsh "niche modelling". A study on niche modelling of saltmarsh plant species (Clarke *et al*, 1993) was carried out to "develop and extend to other saltmarsh plant species a predictive mathematical model of vertical range (or 'niche') of the saltmarsh grass *Spartina anglica*, and to incorporate different components of tidal submergence into such models, increasing their utility in non-sinusoidal tidal regimes". Tidal parameters were derived every 20 minutes throughout the year and eight other saltmarsh species were measured for elevation limits (*Puccinellia maritima, Halimione portulacoides, Festuca rubra, Elymus pycananthus, Juncus gerardii, Juncus maritimus, Scirpus maritimus* and *Phragmites australis*). Table A2.1 shows an example output from the niche approach.

Table A2.1 *Example of number of hours uncovered for* Puccinellia *spp (from ETSU 1993)*

	Hours uncovered in autumn (Sep–Nov) for upper limit		Hours uncovered in autumn (Sep–Nov) for lower limit	
	Max	Min	Max	Min
Puccinellia	971 hours	812 hours	949 hours	682 hours

The main conclusions of this study are as follows, and are useful when designing a managed realignment site with saltmarsh.

1 Mean high water neap (MHWN) tide level produced the most effective method to predict the *species* limit using one standard tidal variable. Between 70 and 96 per cent of the variance in the upper and lower limits, depending on the species, could be accounted for using a simple linear regression with MHWN. However, *Spartina anglica* has a better correlation with mean high water spring (MHWS) tide level.

2 The size of an estuary has a significant effect on the elevation limits. In larger estuaries, the limits of species are farther up the shore than would be predicted by MHWN alone. This is because of the generally greater degree of exposure to wind and wave action, increased velocity of flows and higher turbidity variation.

3 Out of all of the submergence parameters, no single parameter best signified the limit of all species. However, the "number of hours uncovered during daylight in the autumn (Sept–Nov)" consistently predicts species limits with relatively little error. The range of hours should be applied to consider species health.

4 *Spartina anglica* was more accurately predicted with a number of hours submerged by more than 50 cm than other species. This reflects the impact of prolonged tidal submergence on photosynthesis and growth.

Spartina has been believed to be an invasive species in some locations, out-competing other vegetation and limiting biodiversity. There was no evidence of *Spartina* restricting the niche of other species in this work, but it was found that *Spartina* limits may be farther up-shore than expected when certain species (such as *Puccinellia*) are present. The competition between species at elevation niches may not be problematic, but it may need to be understood to avoid false expectation of the vegetation growth on a site and hence how it might look and function. The elevational niche of *Spartina*, in the example below, clearly extends below that of *Puccinellia* by an average of 68 cm, an elevational niche that can include very extensive areas of shallow mudflats (Gray and Mogg, 2001). (Note: in some locations *Spartina* is considered undesirable, as it is an introduced, rather than a natural, species. This should be clarified with the local conservation agency.)

Box A2.2   *Niche competition between species*

> **Spartina**
>
> Upper limit (m) = 0.12 + 1.80 MHWN (R2=89.5)
>
> Lower limit (m) = 0.23 + 1.39 MHWN (R2= 92.3)
>
> **Puccinellia**
>
> Upper limit (m) = 0.21 + 1.71 MHWN (R2= 96.0)
>
> Lower limit (m) = 0.55 + 1.44 MHWN (R2= 95.5)
>
> (Gray and Mogg, 2001)

It was also found that latitude and wave exposure may influence the niche level and so the niche model was further developed. The regression equation that best described the upper limit of *Spartina* was:

Upper limit (m) = 4.74 + 0.483R + 0.068(F) – 0.099(L)

(R^2= 90.2, SE = 0.50)

Where:

R= spring tide range (m)

F = fetch in the direction of the transect (km)

L= latitude (decimal °N) (Gray and Mogg, 2001).

Therefore, saltmarshes with a shorter fetch are likely to have *Spartina* extending farther upshore than would be predicted for tidal range alone.

A2.1.2   Site gradient

Analysis of saltmarsh vegetation has found that slope is important in determining the type of vegetation on sites (Burd, 1995). This is confirmed by studies carried out on marsh restoration schemes in the United States, where slopes of 0–2 per cent were recommended (Zedler, 1984), although later work suggested that 6–7 per cent was feasible (Knutson and Allen, 1990). On historical sites the measured slopes were generally less than 0.1 per cent, with very uniform vegetation types resulting. The complex creek and pan systems that typify natural saltmarshes contribute to micro-topographical and geomorphological variability that also support different plant assemblages. Thus there is a wide range of slopes that will form saltmarsh, and variability will help in achieving a more diverse vegetation cover.

A2.1.3   Wave conditions

Successfully establishing saltmarsh requires a degree of protection from wave action. In essence there is a critical level of wave energy that will erode a saltmarsh. The leading edge of a marsh may (re-)erode and (re-)advance on an annual basis depending on seasonality of the wave climate. The established vegetation, however, damps down wave energy, protecting inner parts of the marsh and inland defences. Wave exposure is less constrained on open coasts than in estuarine locations, so wave decay across the intertidal profile will need to be assessed.

Wave attenuation over saltmarshes

Box A2.3 *Key references for wave attenuation over saltmarshes*

> Knutson, P L, Brochu, R A, Seelig, W N and Inskeep, M (1982). "Wave damping in *Spartina alterniflora* marshes", *Wetlands*, vol 2, pp 87–104
>
> Kobayashi, N, Raichle, A W and Asano, T (1993). "Wave attenuation by vegetation", *J waterway, port, coastal and ocean engineering*, vol 119, no 1, pp 30–48
>
> Möller I, Spencer T, French J R, Leggett D J and Dixon M (1999). "Wave transformation over salt marshes: a field and numerical modelling study from North Norfolk, England", *Estuarine, coastal and shelf science*, vol 49, no 3, pp 411–426
>
> Spencer, T and Möller, I (1996). *Wave dissipation over salt marsh surfaces*. Report OI/569/3/A, Environment Agency Anglian Region
>
> Spencer, T, Möller, I and French, J R (2003). *Wave attenuation over saltmarshes*. R&D Project W5B-022, Environment Agency, Bristol
>
> Whitehouse, R J S, Soulsby, R L, Roberts, W and Mitchener, H J (2000). *Dynamics of estuarine muds: a manual for practical applications*. Thomas Telford, London

One of the applicable wave attenuation formulas for saltmarshes (vegetation damping) is the Modified Dean Model (original Dean Model (1979) modified by Knutson *et al*, (1982)). This applies parameters on cordgrass (*Spartina* spp) size and density and introduces an additional coefficient (C_P) as an empirical adjustment coefficient to account for stem deflection.

The principal variables affecting wave dissipation were considered by Dean (1979) to be:

- the height of the wave approaching the marsh
- the width of marsh through which the wave propagates (shore normal or parallel)
- the water depth
- the diameter of the plants (stem or foliage).

The ratio of incident wave height (H_1 in m) seaward of a stand of marsh grass, and the wave height (H_2 in m) landward of a stand of marsh grass are related as in Equation 1.1.

Modified Dean Model

$$\frac{H_2}{H_1} = \frac{1}{1 + A_1 H_1 \Delta x}$$

(A2.1)

Where:

$$A_1 = \frac{C_P C_D D}{3\pi S^2 d}$$

C_P = stem deflection coefficient

C_D = drag coefficient (-)

D = grass stem diameter (m)

S = average spacing of grass stems (m)

d = water depth (m)

Δx = cross-shore width of the stand of grass through which waves propagate (m).

This formula relates to cordgrass (*Spartina* spp) and it should not be assumed that this is the only vegetation type to consider. Many areas may expect cordgrass or reedbeds (*Scirpus* spp and *Pharagmites* spp) to establish, however, and this formula can then be directly applied.

In a field study by Spencer and Möller (1996) wave energy reductions of 2–55 per cent (mean 26 per cent) were observed across the sandflat section of the intertidal profile, while wave energy reductions of 47–100 per cent (mean 79 per cent) were observed across vegetated saltmarsh. A reduction in water depth from sandflats to the saltmarshes (along a transect) does not explain the increase in wave attenuation over the saltmarsh. Additional processes are also operating, possibly associated with the surface characteristics.

Wave energy reduction in shallow water results from many individual factors, including energy lost through viscous friction, percolation into the substrate and surface friction at the seabed boundary layer. On many saltmarshes the leading (seaward) edge has a greater surface roughness as a result of erosion/accretion cycles and interaction with vegetation. This is particularly apparent where a saltmarsh is in transgression under rising sea levels (or other change towards erosion). In addition, wave reflection, diffraction and refraction can lead to the redistribution of wave energy along the wave crest, which may bring about very localised wave energy reduction (USACE, 1984).

Spencer and Möller (1996) developed a one-dimensional numerical model to quantify the combined effect of shoaling, viscous friction and percolation on wave energy dissipation across a saltmarsh and mudflat. This model was expressed in the form of the equation below.

Spencer and Möller Model

$$H_2 = H_1.K_s.K_v.K_p.K_f \qquad\qquad (A2.2)$$

Where:

H_1 = incident wave height (m)

H_2 = wave height (m) landward of a stand of marsh grass of width Δx (m)

K_s = shoaling coefficient

K_v = viscous friction coefficient

K_p = percolation coefficient

K_f = bed friction coefficient

Wave shoaling due to a decrease in water depth causes wave heights to increase and waves to steepen. Assuming the energy transmitted shorewards between the two rays remains constant, the shoaling coefficient, K_s is calculated according to USACE (1984) as:

$$Ks = \sqrt{\frac{n_1 c_1}{n_2 c_2}}$$

Where c is the wave celerity, the subscripts indicate station number in a shoreward direction and

$$n = \frac{1}{2}\left(1 + \frac{4\pi h/L}{\sinh(4\pi h/L)}\right)$$

With regards to viscous friction, the effect of the viscous boundary layer at the seabed-water interface over a smooth impermeable surface can be quantified using the following equation (Sleath, 1984):

$$Kv = e^{-a_1\Delta x} \approx \left[1 + a_1\Delta x\right]^{-1}$$

where a_1 can be approximated by:

$$a_1 = \frac{(2\pi L)^2}{\left(\frac{4\pi h}{L} + \sinh\left(\frac{4\pi h}{L}\right)\right)\sqrt{\frac{\pi}{Tv_k}}}$$

Where L is the wave length, T is the wave period, h is water depth and v_k is the kinetic viscosity of water (in m^2s^{-1}).

The attenuation effect of friction caused by seabed roughness is represented by a friction factor f. The friction decay factor, K_f, can be estimated by:

$$K_f = \left[1 + \frac{64\pi^3}{3g^2}\frac{fH_1\Delta x}{h^2}\frac{h^2}{T^4}\frac{K^2}{\sinh^3(2\pi h/L)}\right]^{-1}$$

Where g is the gravitational constant and the expression in brackets is the friction coefficient, K_f and f, the friction factor is defined by:

$$f = \frac{\tau_0}{\frac{1}{2}pa_u^2}$$

The percolation decay factor (K_p) depends on the permeability of the surface. Sleath's (1984) equation for K_p can be used:

$$K_p = \left[1 + a_2\Delta x\right]^{-1}$$

It should be noted that this equation assumes negligible bed movement.

Spencer and Möller (1996) and Möller et al (2002) also suggested that the dissipation across an intertidal profile is not linear, but decays exponentially over the marsh surface in the form of Equation 3 (see also Kobayashi el al, 1993).

Spencer and Möller Exponential Decay Model

$$\frac{H_2}{H_1} = e^{-a.\Delta x} \tag{A2.3}$$

Equation 3 was solved by Möller (Möller et al, 1999) to determine that the observed wave height reductions of 63 per cent at Stiffkey, North Norfolk, occurred within the first 100 m of marsh width. Further work on the Dengie Peninsula, Essex, with saltmarshes fronted by mudflats, concluded that there was little seasonal difference in significant wave height attenuation between the mudflat and over the landward sections of the site. More pronounced seasonal differences in wave attenuation were observed at the marsh edge, however. Furthermore, a consistent decrease in attenuation was observed from autumn through winter and into spring. This is for an open coast environment with a wide (c 3 km) intertidal profile of sandy mud/muddy sand.

APPENDICES

With regard to spatial variability, the results showed that, although the dissipating effect of the marsh surface on incident waves appeared greatest in the first 10–20 m of permanent vegetation, increased attenuation occurred up to several 10s of metres inland of the marsh edge. Little to no additional attenuation took place landwards of c 70 m inland of the marsh edge. In the case of a cliffed marsh edge, wave heights increased at the edge but were dissipated rapidly in the first 10–20 m landward of the cliff. It is important to note, however, that the natural cliff topography was considerably more complex than that simulated in previous model experiments. As a result, one would expect a more complex pattern of wave transformation processes to occur under natural conditions (Möller *et al*, 1999). The results provide empirical support, both for maintaining saltmarshes in front of existing sea defence lines and for creating new saltmarsh as part of coastal set-back/shoreline realignment schemes.

A2.1.4   **Soil type**

Clay or clayey loam soils are considered to facilitate the development of the saltmarsh habitat more rapidly than gravel, sandy or alluvial soils, although all soil types are acceptable and have different plant assemblages. Experience in re-creating saltmarsh on sandy dredge spoils in San Diego Bay found that although plants re-established, a lack of nitrogen limited optimum growth and the full ecological potential of the site was not attained (Zedler and Adam, 2002). In most managed realignment sites the existing landward soil will have been enclosed and reclaimed from the sea in the past; the soil chemistry, however, will have been altered in this process and will need time to readjust for natural saltmarsh to develop. The soil of drained saltmarshes can lead to sulphide accumulation and the formation of acid-sulphate soils when intertidal conditions are restored (Zedler and Adam, 2002).

The period since enclosure took place is relevant (RSPB, 2000). This is because site elevation generally reduces over decades as soils settle and dry out following enclosure, so the probability of successful saltmarsh restoration is greatest where the period since enclosure has been shortest.

A2.1.5   **Water quality**

Other factors that may influence intertidal habitat creation include water quality parameters such as salinity, nutrient supply, turbidity and the presence of toxic compounds (Burd, 1995).

Freshwater may influence any contamination in saltmarshes (by modifying the chemistry) and low salinity may lead to invasion and out-competing by plants such as reeds (*Phragmites australis*) and reedmace (*Typha* spp). Changes in freshwater inputs to estuaries, whether as a result of natural shifts or human intervention within river catchments, can lead to change in saltmarsh but may not be considered to be problematic.

With the exception of salinity, there is relatively little information on the significance of chemical factors. A body of work on saltmarshes and various chemical compounds (including agro-chemicals) has been undertaken for the Environment Agency by Imperial College (for example, Imperial College, 1992; Meakins *et al*, 1995; Leggett *et al*, 1995). Turbidity may be expected to reduce photosynthesis, but as plants are exposed to the air for the majority of the time this may not be particularly important, indeed this is part of the factors that limit the extent of saltmarsh.

A2.1.6 Biological factors

Most sites potentially suitable for managed realignment are close to existing saltmarsh that can naturally provide the necessary seeds. Seeding or planting of saltmarsh to increase the likelihood of successful vegetation regeneration is not usually necessary. It has been suggested that plant-eaters (such as rabbits or geese) may occasionally limit the growth of saltmarsh plants.

A2.2 INTERTIDAL FLATS

Box A2.4 *Key references for intertidal flats*

> Atkinson, P W, Crooks, S, Grant, A and Rehfisch, M M (2001). *The success of creation and restoration schemes in producing intertidal habitat suitable for waterbirds*. Research Report 425, English Nature, Peterborough
>
> Joint Nature Conservation Committee (1996). *Guidelines for the selection of biological SSSIs; inter-tidal marine habitats and saline lagoons*. JNCC, Peterborough
>
> Pethick, J S (1984). *An introduction to coastal geomorphology*. Edward Arnold, London, 260 pp
>
> Roberts, W (1992). *Fluidisation of mud by waves: development of a mathematical model of fluid mud in the coastal zone*. Report SR 296, HR Wallingford, Wallingford
>
> US Army Corps of Engineers (USACE) (1984). *Shore protection manual*. Engineering Research and Development Centre, Coastal and Hydraulics Laboratory, Vicksburg, MS
>
> Whitehouse, R J S, Soulsby, R L, Roberts, W and Mitchener, H J (2000). *Dynamics of estuarine muds: a manual for practical applications*. Thomas Telford, London

Intertidal habitats form in lower-energy coastal environments (including wide, shallow, environments), in estuaries and behind spits or bars where they are sheltered. Mudflats consist of fine sediment material such as silts and clay, and have a high organic content. Mudflats often appear in a natural sequence of habitats and can be found as a transitional zone between saltmarsh and sub-tidal channels (English Nature *et al*, 2003). Mudflat sediments are mostly transported in suspension, and the sediment is deposited when the velocity of the tide is low (ie < 0.1 cm/s). Sand and gravel may be deposited under higher flows (and exist where there is a greater disturbance due to wave action). Figure A2.3 illustrates the relationship of sediment erosion, transport and deposition with respect to water velocity, assuming this is applied to the size of sediment. It is important to note that actual measurements may be at some distance above the bed and simple ("law of the wall") transformations may not be appropriate.

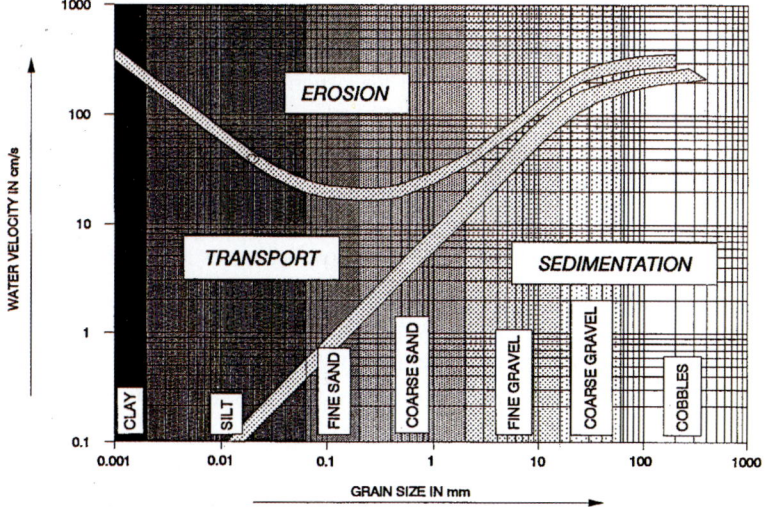

Figure A2.3 *Erosion, transport and sedimentation of different sediment sizes*

As more sediment (and/or organic matter) is deposited, the level of the bed will rise so that eventually it becomes exposed at low tide for long enough to be colonised by species tolerant to submergence and salinity. The first species to colonise are usually benthic microalgae, especially epipelic diatoms. These minute organisms secrete large amounts of mucus, which help bind the finer sediment together and increase surface stability. This process is more apparent in mudflats, but it can also benefit very fine sand in quiescent conditions. Algae are tolerant to frequent tidal submersion and high salinity, but some species (*Enteromorpha* spp, *Ulva lactuca*, *Fucus*) may be sensitive to sediment stability and only tolerate low accretion rates. *Zostera* spp (eelgrass) is typically the next species to colonise an intertidal area, thus increasing friction of the intertidal flat and usually adding to the amount of sediment deposited. This raises the elevation further, eventually enabling colonisation by saltmarsh species such as *Salicornia* (glasswort).

A2.3   SAND DUNES

Box A2.5   *Key references for sand dunes*

> Brooks, A and Agate, E (2000). *Sand dunes: a practical handbook*, 4th edn, British Trust for Conservation Volunteers, Wallingford
>
> English Nature, Environment Agency, Defra, LIFE and NERC (2003). *Living with the sea. Coastal habitat creation: towards good practice* [www11]. English Nature, Peterborough
>
> Nordstrom, K F, Psuty, N P and Carter, R W G (eds) (1990). *Coastal dunes: processes and morphology*. J Wiley & Sons, Chichester
>
> Ranwell, D S and Boar, R (1986). *Coastal dune management guide*. Institute of Terrestrial Ecology (NERC), Huntingdon
>
> Scottish Natural Heritage (2000). *Beach dunes: a guide to managing coastal erosion in beach/dune systems*. Scottish Natural Heritage, Redgorton, Perth
>
> Van der Meulen, F, Jungerius, P D and Visser, J H (eds) (1989). *Perspectives in coastal dune management*. SPB Academic Publishing, The Hague, pp 207–216

In the absence of vegetation, sandflats may function as a sediment supply to dune ridge systems if they are wide enough and if the prevailing wind allows. Dunes will form naturally where a reduction in wind velocity causes sediment to be dropped or where a topographical change allows sediment to become trapped and vegetation to establish. Sand dunes can attain great height and width, but they may also constitute a dynamic environment in which erosion, transport and re-forming take place in response to annual or less frequent changes in conditions. Dunes, especially foredunes or embryonic dunes, may be eroded away under winter (storm) wave action and re-form in summer (swell) conditions. Indeed, the maintenance of this type of habitat may depend on such changes taking place. In this way initial dune formation is linked to beach processes and might be seen as part of the beach energy dissipation system, and a sediment store for the upper part of the profile (www13, see Futurecoast CD).

A2.3.1   Sand dunes and managed realignment

Although sand dunes have not commonly been considered for managed realignment, the concept should not be ruled out. The principle for sand dune managed realignment is the same as that for shingle ridges (see below), ie to facilitate the restoration of natural processes or to manage the dune ridge to a new landward position. The sand could be either artificially (mechanically) moved landwards or the dune modified to allow aeolian (wind) transport of sand landwards. This would reduce the pressure of coastal squeeze and erosion of the front face of dunes by rising sea level and/or allow the full range of dune landforms to evolve. The retreat of sand dunes could produce a more sustainable coastline by widening the intertidal (beach).

Vegetation (such as marram grasses) may need to be planted if the whole dune system has been disturbed and allowed to move inland or to limit the landward transgression of the dune system (perhaps including the use of bushes and trees).

It may be desirable to allow dunes to move to a more landward position as sea levels rise and the dunes erode. This position can be designed by creating an inland shelter-belt or increasing surface roughness (for example, using brushwood). The use of dune management techniques in the past, however, may have created a situation where the dune is not self-sustaining and cannot respond unaided to such changes in environmental conditions. Erosion of dunes does liberate sediment into the intertidal, and may be reworked to a new landward, dune position. If energy levels have generally increased, however, sediment may be moved offshore or lost alongshore from the dune system. Also, severe dune erosion may alter the shape of the dune such that sediment transport to landward by wind action is impaired.

A2.3.2   Formation and development

Dune habitats can be of high ecological interest because of the rare species of flora and fauna they contain, and protection of these habitats is important.

Figure A2.4   *Sand dune habitat*

Woodhouse (1982) classified three stages in dune succession: pioneer, intermediate and mature. The pioneer stage refers to the formation of embryonic dunes and foredunes, intermediate stages cover the creation and "roll-over" of dune ridges, and the mature stage refers to old dune ridges where transition to terrestrial vegetation may occur.

Sand dunes form when sand is blown up the shoreline during low tide by saltation ("bouncing") and surface creep. The sand starts to accumulate around debris or vegetation, causing more sediment to be deposited until an embryo dune forms. This then accumulates further and gradually vegetation (such as marram grass) begins to colonise. Vegetation causes a greater surface friction and a wind shear that leaves a zone of low wind velocity near to ground level (a Z° layer), causing sand to be deposited and not re-entrained.

As the foredune height increases towards a self-limiting level it comes under greater pressure from wind velocities. This results in the movement of sand from the crest and deposition on the landward lee slope. The entire dune moves landwards (the roll-over process) (Pethick, 1984; Carter, 1990). This allows further dunes to develop on the shore and starts the formation of dune ridges.

APPENDICES

Dune slacks develop where the water level is higher than the sand, which is caused by erosion by wind at the bottom of the lee slope after sand is deposited on the crest and top of the lee slope. This leads to the development of a wet hollow known as a swale (dune slack). As the dunes mature and are rolled back they eventually become lower and less parallel to the shoreline. This coincides with blowouts (areas of erosion), causing topographical variation.

A2.3.3   Erosion of sand dunes

Figure A2.5 shows the effects of erosion and deposition of the dune profiles.

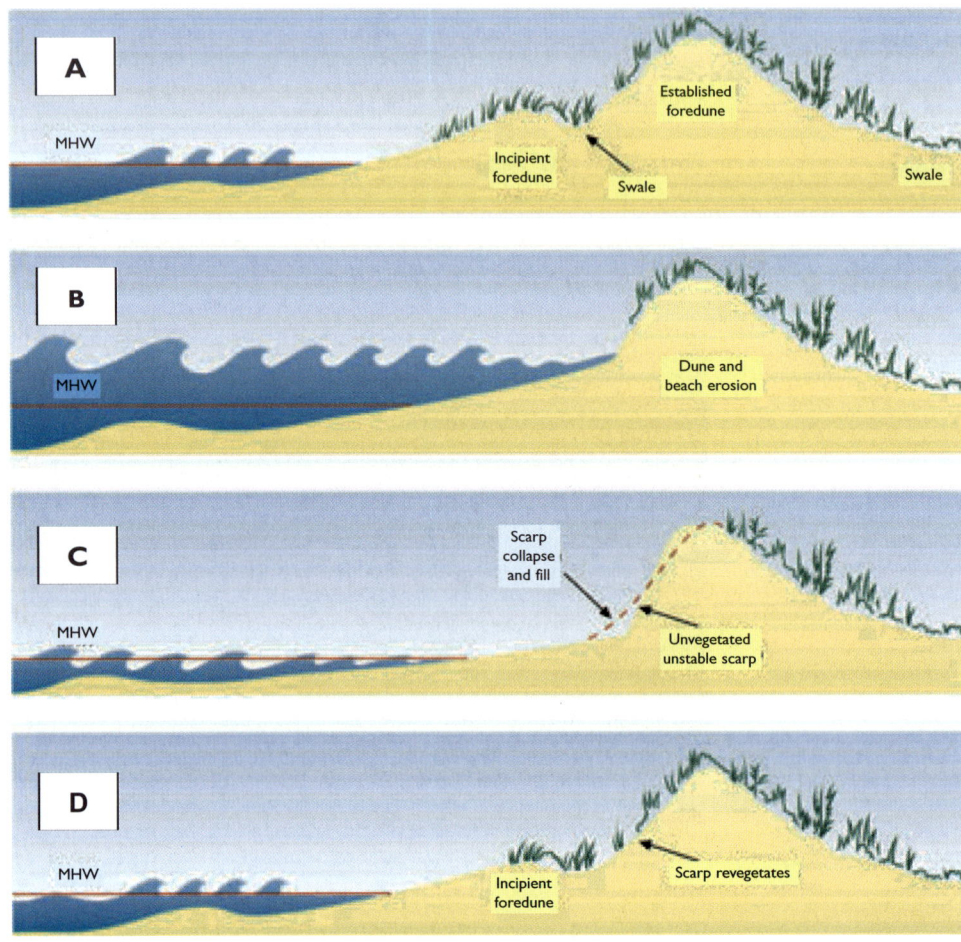

Figure A2.5   *Dune-building processes and the effect of sand build-up (accretion) and erosion. Image copyright Hesp, courtesy New Zealand Government*

   A   Once a foredune system has developed, vegetation colonises, creating a fairly stable habitat.

   B   During storm surges, the sea level exceeds the MHW level. This can cause erosion of the foredunes. A steep or cliffed sand dune is formed at the seaward edge (scarp slope) and vegetation is lost.

   C   The unstable scarp slope begins to slump and form a gentler scarp slope and becomes more stable.

   D   The stabilised slope eventually begins to recolonise and a new embryo dune develops in front of the foredune.

Analysis from Vasseur and Héquette (2000) suggests that there has been an increase in storm magnitude and frequency (between 1968 and 1988), and that there is a strong relationship between dune-front erosion and frequency of storm surge conditions. Also, with increasing risks of sea level rise, the probability that coastal dunes will be breached by storm waves is likely to increase. Thus the risk of erosion and, therefore, the risk of flooding of coastal lowlands behind dunes are likely to increase.

The adaptation of sand beach profiles (included those backed by dunes) to sea level rise should be allowed for. The Bruun Rule provides a way of estimating change and conceptually understanding how and where change might occur across an intertidal profile.

Box A2.6   *The Bruun Rule*

The Bruun Rule can be used to indicate the relation between beach retreat and sea level rise. It is based upon the principle of the beach equilibrium profile: for a given wave climate and grain size, the beach will attain a predictable profile equilibrium. If sea level rise occurs, it is predicted that erosion will occur at the shallow end of the beach profile and deposition will take place offshore on the seaward end of the profile. Thus the equilibrium profile is maintained (Simm, 1996). The following equation from Bruun calculates the water depth.

$$h = Ax^{2/3}$$

Where:

h = water depth

x = horizontal distance from shore

A = constant dependent upon grain size and on the length units used for x and h

The shoreline retreat is approximated by:

$$\Delta y = l\Delta s\, /\, ha$$

Where:

Δy = shoreline recession

$l\Delta s$ = theoretical deposition required to re-adjust the profile
    (l = profile length, Δs = sea level rise)

ha = depth of exchange of material

The profile is assumed to be in balance, and therefore, the only way in which the material for deposition can be obtained is by a shoreward movement.

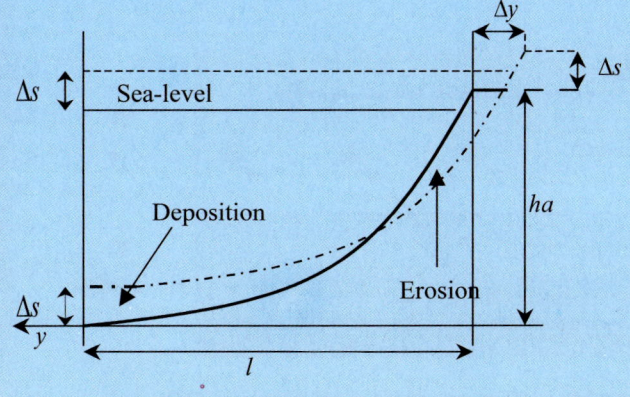

A2.4 SHINGLE RIDGES

Box A2.7 *Key references for shingle ridges*

> Doody, P and Randall, R (2003). *A guide to the management and restoration of coastal vegetated shingle*. English Nature, Peterborough
>
> English Nature, Environment Agency, Defra, LIFE and NERC (2003). *Living with the sea. Coastal habitat creation: towards good practice* [www[11]]. English Nature, Peterborough

Shingle ridges form when shingle sediment is deposited by wave action on the beach, forming a ridge. Storm waves enable sediment to be deposited higher up the ridge and not to return seawards. If storms are of significant magnitude they will breach this ridge or overwash it, causing the feature to migrate inland or enabling a sequence of ridges to form. The behaviour of shingle ridges is a balance between sediment supply and wave/tide forcing.

A2.4.1 Evolution of shingle ridges

> *Shingle is the term applied to sediments larger in diameter than sand but smaller than boulders (generally between > 2 mm and < 200 mm). At a regional scale, the type of rock determines shingle availability and durability. Hard materials such as flint, chert, granite, quartzite, and some metamorphic materials survive much longer at the size range than sandstones, limestones or shells.*

(Doody and Randall, 2003).

The distribution of shingle ridges is restricted to areas of higher wave energy and the shingle resource is widely distributed in Great Britain (Doody and Randall, 2003). Shingle beaches and structures have evolved over geological time, and in the UK the majority of the shingle has derived from glacial deposits and the erosion of chalk cliffs (exposing flint) after the last Ice Age. This material has been transported along the coast by longshore drift.

Longshore drift occurs when there is an oblique wave approach to the shoreline (Komar, 1998) (see figure below).

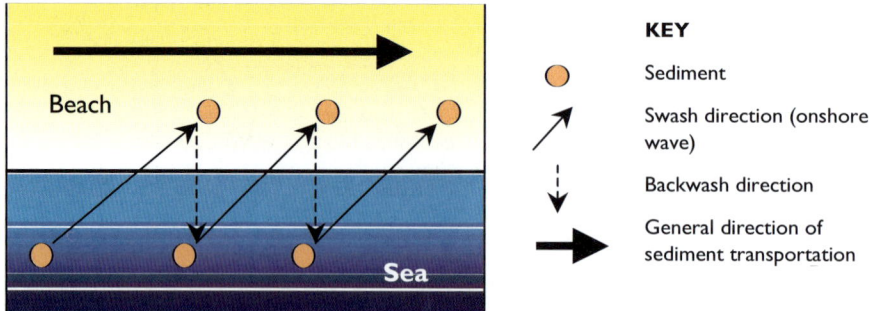

Figure A2.6 *Diagrammatic view of longshore drift*

A2.4.2 Shingle ridge landforms

Where there is a greater sediment input into the system than sediment output, longshore drift can create extensive shingle deposits such as forelands, spits and nesses. These deposits tend to occur at the terminus of a sediment supply (eg at an estuary mouth) and form sediment sinks (ie sediment is deposited and stored).

These features tend to be aligned to the net longshore drift, but they may be modified and aligned, in part, to wave swash processes. Drift alignment thus happens with sufficient sediment supply. The sediment input of many ridges in the UK today is much reduced in comparison with historic levels. This is explained by the relict nature of glacial gravel deposits and by the reduced input of eroded cliff sediment as a result of shore protection and control structures.

Swash-aligned shingle landforms are aligned along the shore in a perpendicular angle to the onshore wave direction. Where there is a lack of sediment coming into the system, "cannibalisation" can occur, causing the ridge to roll landwards by erosion of the seaward face and transport landward (or seawards, in which case the ridge decays). Landform mobility is caused by tide/wave overwashing, removing sediment from seaward and moving it landward or offshore. Overwashing can help build up barrier crests and width, and as this process continues, washover fans can develop. Vegetation can grow on the crest, but is usually lost under storm surges; this is an important part of the ecological functioning of the system.

Other significant impacts on shingle (English Nature *et al*, 2003) can be caused by:

- inappropriate offshore dredging or extraction, where this depletes the sediment available to the system or causes direct drawdown of the shingle (this is less likely to occur now than historically, as impact assessments are undertaken)

- habitat loss from construction on shingle surfaces

- sea defence and coastal protection works

- recreation

- onshore aggregate extraction.

Shingle ridges will retreat, especially in the south of England, where sea level rise and isostatic readjustment combine to cause greater sea level changes compared with, say, Scotland. In Scotland, shingle ridges do not seem to retreat as much because of rising land levels from isostatic rebound, so change might be caused by increased wave energy (from greater storm activity).

A2.4.3   Managing shingle ridge profiles

In the UK, many shingle ridges:

- provide protection for other coastal habitats (especially wetlands)

- are unstable and trying to move towards swash alignment as drift alignment is decreasing

- are suffering from reduced sediment supply

- are being artificially held, by management, in unsustainable positions

- have experienced, or will experience, cannibalisation of sediment

- will eventually erode and become sediment sources

- may be artificially stabilised through the importation of large volumes of sediment to maintain their present flood defence standard, but this is likely to become increasingly uneconomic.

Managed shingle ridge profiles can be much steeper than unmanaged profiles. This is a result of moving sediment (mechanically) upwards to attain high crest levels and/or seawards to maintain the existing defence alignment. This process pins the ridge into a region of greater wave energy, rather than allowing the natural evolution and migration of the ridge. This is particularly relevant where the ridge is no longer drift-aligned and is now swash-aligned or subject to cannibalisation of sediments. Steeper profiles decrease the flood risk as the height is greater (therefore, the storms are less likely to overwash the ridge), but steeper profiles can increase the risk of catastrophic breach. Unmanaged profiles are usually much wider, but not as high. This generally reduces catastrophic breach, but can increase risk of overwashing. As the width is greater in a natural ridge, wave energy can be dissipated across a wider surface. If a managed profile is left to become natural, then it would be rolled back initially at an enhanced rate compared with a totally natural ridge, but could then be more stable in a new equilibrium position.

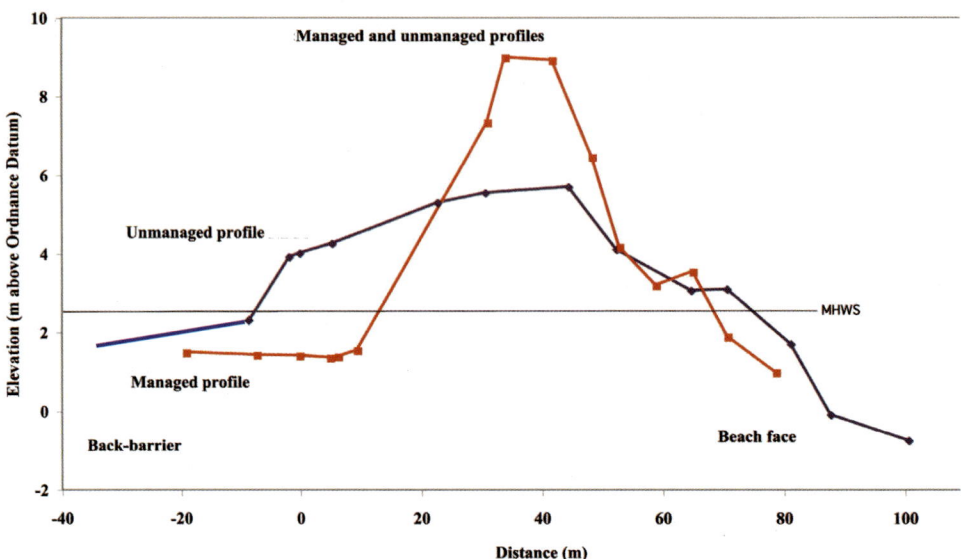

Figure A2.7   *Example of managed and unmanaged beach profiles (source: Environment Agency, Anglian Region, Shoreline Monitoring Programme)*

A2.4.4   Shingle ridges and managed realignment

If a shingle ridge is being realigned or restored, then it is advisable to try to reproduce a more natural (wider) profile. This creates a better energy absorber, however, storm surges may still overtop the ridge and thus new landward defences may be required. Alternatively, a ridge can be moved inland and maintained in a new position that is less exposed to storm activity as a result of a wider, fronting, intertidal beach. This may allow a semi-natural profile to be used that is wider and lower than the existing managed ridge.

In long-term ridge managed realignment schemes, caution should be used when assessing how stable a ridge might be (either as a flood defence or as a habitat). Where sea-level is predicted to rise, more roll-back of the ridge may occur. Also, an ongoing reduction in the sediment supply may require maintenance of the shingle barriers volume (by importing material), although this is not always a feasible option.

Figure A2.8 *Managed shingle ridge*

A2.4.5 **Shingle habitats**

Possible approaches to the restoration of vegetated shingle (Walmsley and Davy, 2001) include:

- allowing natural regeneration
- using the natural soil seed bank
- sowing seed *in situ*
- planting container-grown plants.

The effect of previous management regimes on the suitability of the substrate will need to be considered, for example the grading of imported (sand and) shingle, past interference with the beach profile and possible compaction or disturbance of existing substrates. A proportion of fine material in shingle substrates greatly improves plant germination and survival, but excessive quantities encourage invasion by weed species (Walmsley, 2002). There has been some research on suitable techniques, which should be reviewed before undertaking shingle restoration (for example, see English Nature *et al*, 2003; Perrow and Davy, 2002).

Vegetated shingle habitats are rare because of a lack of nutrient-rich beds and their limited geographical extent. In the design of a shingle ridge realignment it is important to consider conditions under which vegetation can colonise. It is also important to take into account factors that may limit the successful establishment of shingle vegetation species, including:

- invasion of undesirable species (bramble, traveller's joy etc). This is especially relevant where finer sediments are mixed with the gravel and usually occurs only when shingle is above the reach of waves. However, some finer material is desirable for vegetation colonisation
- sea level rise causing an increase in storms, thus eroding habitats on top of shingle ridges and roll-over (although recovery may ensue after the storm)
- vehicular and pedestrian impact on shingle (although this might be controlled to some extent).

A2.5 SALINE LAGOONS

Box A2.8   *Key references for saline lagoons*

> Bamber, R N, Gilliland, P M and Shardlow, M E A (2001). *Saline lagoons: a guide to their management and creation*. Saline Lagoon Working Group, English Nature, Peterborough
>
> English Nature, Environment Agency, Defra, LIFE and NERC (2003). *Living with the sea. Coastal habitat creation: towards good practice* [www[11]]. English Nature, Peterborough

Saline lagoons develop where a barrier allows saline water to be retained. Where these are natural they are usually small and ephemeral as beach roll-over or changes in salinity reduce the brackish conditions in the water. Larger lagoons also develop in tidal inlets sheltered from storm waves by enclosing spits and bars. They are usually defined by their physical situation and salinity. There are five main physical types: isolated lagoons, percolation lagoons, silled lagoons, sluiced lagoons and lagoonal inlets. The definition is based on their physiography, ie by the nature of the separating barrier (Bamber *et al*, 2001).

Water quality and salinity levels are critical to their value for the specialist marine flora and fauna they support. Management is usually concerned with ensuring that both elements are appropriate to the interest present on the site. There are differences in the requirements, but the preference of most marine-related species is for salinities approaching that found in the sea, ie 35 ppt, and with some exchange of water. Low sedimentation rates also allow lagoons to continue in existence, but most natural lagoons will eventually infill unless they are large enough to prevent this through disturbance of the sediment by waves.

Some 30–40 lagoons are estimated to have been lost in England in the 1980s. These losses, which can be exacerbated by natural change, continue to affect the remaining sites today and provide powerful incentives for restoration. Other important factors include the size and shape of the lagoon, the substratum type and the need to control encroaching vegetation such as common reed. It should also be recognised that the forces causing natural loss of lagoons (from coastal erosion) may also be drivers for realignment. In some cases, lagoons may be adversely affected by realignment, in which case it may well be necessary to re-create this type of habitat within the scheme design.

A2.6 INTERTIDAL REEDBEDS

Box A2.9   *Key references for intertidal reedbeds*

> English Nature, Environment Agency, Defra, LIFE and NERC (2003). *Living with the sea. Coastal habitat creation: towards good practice* [www[11]]. English Nature, Peterborough
>
> Hawke, C J and José, P V (1996). *Reedbed management for commercial and wildlife interests*. reserve Management Sheet no 4, RSPB, Sandy
>
> Nuttall, P M, Boon, A G and Rowell, M R (1998). *Review of the design and management of constructed wetlands*. Report 180, CIRIA, London

Reedbeds in the UK and most of Europe are dominated by *Phragmites australis*. They occur at the margins of tidal land where brackish to freshwater transitions occur, and in freshwater pools. *Phragmites* can quickly invade shallow open water and is essentially a primary coloniser.

Reedbeds (and the species that rely upon them) are a rare habitat. Drainage (and enclosure of estuarine reedbeds) has reduced their area considerably.

A2.6.1 Creating reedbed

The first stage of designing a realignment to create reedbed is to profile the substrate to create the desired shape and levels. A network of ditches and areas of open water can be desirable, with sloping ditches and banks providing variety in the habitat and ditches providing habitat in their own right. Water management infrastructures, such as bunds, ditches, dams and sluices, can be installed. Introducing some slightly raised areas, to create variations in water depth in the reedbed, will enhance very flat sites. Surplus soil from excavating ditches can be used to create areas of raised ground that are less prone to flood.

Reedbeds created by land-forming operations where an aim is to provide commercial reed for harvesting may not have as much variation in water depth as is optimum for developing a range of nature conservation interests.

The manipulation of water levels and periods of flooding requires suitable structures to be installed in most areas where the reedbed is of nature conservation and/or commercial importance. An important exception might be extensive tidal marshes at the upper end of an estuary with a fluvial input to it. In these cases, reedbeds may form naturally. A variety of structures can be used including (English Nature *et al*, 2003 [www[11]]):

- bunds (low earth banks designed to maintain water levels above the surrounding land)
- dams (help isolate different sections of reedbed as an aid to rotational management)
- sluices (permit the controlled flow of water into or across a site)
- also important is the storage of water and the use of drainage ditches to facilitate the movement of water through the site.

Once the landform has been designed, together with the appropriate water management structures, reed can be introduced to the site. This may occur naturally but is more likely to require planting. Considerations of importance when introducing reed plants include (English Nature *et al*, 2003 [www[11]]):

- bed preparation (where competing vegetation and water levels must be considered)
- opportunities for obtaining plants or seeds derived from local sources
- methods of establishing seed viability and assessing collection needs
- using the most appropriate method for sowing seed
- the use of pot-grown plants (often more reliable than seed sown directly on site)
- spreading rhizomes at a lower density using existing reeds (ie redistributing existing plants to cover a larger site).

A3 Physical processes and their influence on design

Physical process parameter/ morphological characteristic		Influence on design	Effect of design	
			Within site	Wider environment
MORPHOLOGY	Hinterland topography and geology	Hinterland topography relative to tidal levels determines the extent of tidal inundation and hence the need, or otherwise, for flood defences to control this. The geology of "high ground" hinterland (together with its exposure to forcing conditions) determines its erosion potential in the absence of flood defences. Hinterland use is also a relevant when identifying the need, or otherwise, for flood defences.	Rising topography, or defences fronting low-lying topography, provides the boundaries to the site and hence partly dictates the size of its tidal prism (which is also dependent on site levels and tidal levels). For tidal prism effects, see Section 5.3	Flood or erosion risk to adjacent hinterland areas could potentially be altered (either reduced or exacerbated dependent on scheme-specific factors).
	Within-site topography and morphological features	Topographic levels of the site relative to tidal levels partly determine the size of its tidal prism (which is also dependent on site area). Low topographic level of the site relative to existing intertidal levels could indicate a tendency for ponding of tidal waters and hence identify drainage problems. Site levels can be altered by removing or importing material. Site gradient determines the direction and rate of "drainage" of tidal waters from the site and influences the diversity of potential habitat creation. Gradient can be altered by removing/ redistributing/importing material to the site. Cross-shore width influences the degree of additional wave attenuation due to bed friction, percolation, shoaling and viscous damping that will be experienced as waves propagate across the site. This may have implications for the crest levels of any required flood defences. The presence of antecedent creek features on previously reclaimed sites, or of land drains, borrow pits etc could influence decisions about the need, or otherwise, for internal creek networks, drainage and the location of any breaches/lowering of the seaward defence. The presence of drainage channels and/or soft unconsolidated geology may hinder landward realignment of structural or natural defences (eg subsidence of realigned embankment or inhibition of continued gravel barrier roll-back).	Site topography partly dictates the size of the site's tidal prism (which is also dependent on site area). For tidal prism effects, see Section 5.3. In coastal environments, flood tide deltas could potentially form if a new tidal inlet is created by breaching the existing defence. Spits could also form at the inlet mouth if there is a sufficient longshore supply of non-cohesive sediment transport. Increased intertidal width increases the amount of wave attenuation across the site. This may have implications in terms of reduced resuspension/ increased deposition of suspended sediment and knock-on effects on habitat formation.	The tidal prism generated by a new tidal inlet could lead to the formation of ebb tide deltas just seaward of its mouth in coastal environments.

APPENDICES

Physical process parameter/ morphological characteristic		Influence on design	Effect of design	
			Within site	Wider environment
MORPHOLOGY	Intertidal and sub-tidal topography and morphology features	The level of the existing intertidal may influence the flow of waters to and from the managed realignment site (ie provide a barrier to all but the highest tides or create ponding on the site by preventing drainage). Presence of creeks across the intertidal may influence the selection of location of breaches or defence lowering so as to maximise the continuity between the environments and minimise potential adverse impacts. Existing banks and channels can provide natural sheltering or focusing of energy. Natural variability in the position of morphological features can change exposure conditions to site (eg channel switching, saltmarsh die-back or erosion). Other seabed and shoreline uses and users (eg navigation channels, shellfisheries).	Level difference between the site and intertidal may affect timing of flow on to or off-site and create a "weiring" effect during the flood tide or a "ponding" effect during the ebb tide if the intertidal is considerably higher than the site.	Exacerbation, cessation or reversal of existing natural tendencies for erosion or deposition. Changes in morphological features (eg intertidal creek formation, widening or siltation).
HYDRODYNAMICS	Tidal levels and range	Determines inundation frequency (in combination with within-site topography and any constraining intertidal features) and hence the need/desire for techniques to control frequency of inundation and/or land regrading.	Partly dictates the size of the site's tidal prism (which is also dependent on site area and tidal levels). For tidal prism effects, see Section 5.3. Location of development of particular habitat zones (also dependent on within-site topography).	Water levels and tidal range throughout the coastal or estuarine system may be influenced if the scheme is particularly large in relation to the system within which it is set, or if the scheme is being implemented with other managed realignment schemes that combine to have a large effect.
	Tidal current velocities	Selection of appropriate methods to minimise changes in overall velocity, or changes in velocities at certain, vulnerable locations. Reinforcement of areas such as the breaches, to prevent erosion occurring.	Re-suspension of newly deposited sediment, prevention or occurrence of sediment deposition, erosion of existing morphology (this is dependent on velocities relative to critical thresholds of appropriate sediment size grades). Breach widening and ingress/ egress channel deepening to accommodate higher velocities. Vulnerability of new defences to erosion.	Changes in flow speeds and directions in the wider system leading to erosion/ deposition. Changes to existing intertidal (eg creek widening to accommodate increased flows). Vulnerability of adjacent features/structures to erosion.
	Tidal asymmetry	Selection of appropriate methods to delay or enhance the inflow/egress of water to/from the site (eg to encourage sedimentation).	Influences whether the site is a net source or sink of sediment.	The flood/ebb dominance of the entire estuarine system could potentially be altered (increased, reduced or switched) by the scheme, which could influence the net import or export of sediment.
	Tidal prism	Site prism determines the volume of water entering and leaving the site, so influences the velocities of flow through breaches or sluices/ over spillways. It can be altered by changing the site area by construction of new seawalls or altering site levels through excavation, regrading or import of material. Discharge controls the size of breaches and/or drainage channels necessary to avoid erosion.	Tidal prism (and scheme design) will increase the tidal current velocities on the site (see "Tidal current velocities" above).	The prism of the entire estuary could be increased, potentially leading to morphological adjustments. In coastal environments, significant tidal prisms of schemes could lead to significant water and sediment exchange with the wider system, causing ebb tidal deltas to form.

Physical process parameter/ morphological characteristic		Influence on design	Effect of design	
			Within site	Wider environment
HYDRODYNAMICS	Sea level rise	Crest levels of defences. Realigned natural gravel and dune features will continue to exhibit a tendency for landward transgression in response to long-term sea level rise. Consider influence on wave propagation under climate change.	Long-term landward transgression of landforms (and habitats). Increased tidal prism of site.	The prism of the entire estuary will increase marginally, but progressively. Bank retreat would enable the long-term transgression of existing intertidal landforms, to the limit of any new alignment.
	Wave action	Determines exposure conditions, and hence the need for wave breaks within or outside the site. Determines the crest level of new sea defences within the scheme. May determine the choice of bank or breach retreat (depending on exposure and desired sedimentation rates/habitat formation). Wave action also influences the degree of protection or armouring required for new sea defences within the scheme.	Re-suspension of deposited sediment or prevention of sediment deposition. Changes to the extent of mudflat and saltmarsh. Increased dissipation over wider intertidal profile (through a combination of percolation, surface roughness, shoaling or viscous damping).	Refocusing of wave energy due to development of ebb tide deltas at new inlet mouth in coastal environments. Potential for altered wave climate either side of the realignment scheme. Potential affects on navigation for small craft.
SEDIMENTS	Sediment composition	Influences degree of consolidation of basement for new defences, realigned barriers etc and hence required crest level.	New flood defences or barriers may sink into soft basement material. Barriers may not roll back (landward transgression) with sea level rise. Influences the colonisation of the site by floral and faunal communities.	Potential for increased suspended sediment concentrations released through increased erosion or re-suspension of material, if sediment is fine-grained and erodible. Also influences the susceptibility of existing intertidal areas to erosion through inflow or egress of water. Potential for altered wave climate either side of the realignment scheme.
	Sediment supply and storage	Anticipated sedimentation rates on the site may influence the habitat formation and hence the degree of wave energy attenuation across the site towards any required flood defences.	Sedimentation or erosion potential within site.	Increases or decreases in available sediment to feed other areas of the natural system.
	Sediment transport (bedload)	Need for bypassing or recycling if bedload material causes blockage of breach or inlet mouth due to transported sediment (either alongshore or cross-shore in or out of site).	Transport into the site could lead to the formation of flood tide deltas and/or spits within new inlet mouths in coastal environments. Could ultimately provide shelter to managed realignment site.	Transport out of the site could lead to formation of ebb tide deltas seaward of new inlet mouths in coastal environments or to interruption of longshore drift to downdrift frontages by deltas, spits or new inlet.
	Sediment transport (suspended load)	Blockage of sluices. May influence or negate the need for the design to enhance the deposition of suspended sediment. Need to know location of other seabed and shoreline uses (eg navigation channels, shellfishery beds).	Possible preferential areas for deposition of available suspended load sediment caused by undulating within-site topography or presence or absence of within-site creek systems.	Increase in suspended sediment concentrations caused by erosion of site or existing intertidal or widening/ deepening of creeks and breaches. "Bulk-loading" of resulting sediment plumes to other seabed or shoreline areas. Sediment smothering of shellfish beds or fisheries' interests.

APPENDICES

A4 Review of scheme implementation, including assessment of physical processes

When undertaking managed realignment, lessons can be learned from other realignment sites. Visiting other sites can help in visualising the outcome and identifying and understanding the issues that are faced. Information and knowledge gained in this way may not be directly applicable to other locations, however. Each site is likely to have its own set of drivers, constraints and differences that need to be carefully considered.

If an existing site is local (within the coastal sub-cell or estuary) it may provide direct data and information that can be applied. Care will still be needed in applying such data, because, for example, water level data can vary significantly throughout an estuary. If it is in a different environment, a significantly different physical setting or has different types of defence then it will be important to interpret the relevance of any information gathered.

Site visits should be undertaken with those who were directly involved to gain the greatest benefit. Depending on the information required, this might include the landowner, lead organisation, partner organisations, consultant and/or researchers. To ascertain information for physical processes, those who undertook the design are likely to be the most knowledgeable and it may be reasonable to cover their costs in providing advice.

A4.1 REVIEW OF SCHEMES IMPLEMENTED/NOT IMPLEMENTED

A4.1.1 Managed realignment schemes that have been implemented

This component of the review identifies examples of best practice and lessons learned relating to scheme design or assessment of scheme impacts from those that have been implemented. The information derived from them has been incorporated directly into the various sections of this design guide. Intentionally, no specific case studies have been mentioned by name elsewhere in this guide so that focus remains on the key elements of best practice and lessons learned rather than specific scheme details. However, an overview of some implemented schemes is presented in Table A4.1.

Table A4.1 *Examples of recent managed realignment schemes implemented*

Site, location and county, and starting information source	Date of scheme and retreat area	Initial comments and findings
Northey Island, Blackwater Estuary, Essex (National Trust and Environment Agency)	1991 0.8 ha	One of the first schemes undertaken as a deliberate realignment by initial lowering of a flood embankment, followed by a full breach. The small site was relatively high in the tidal frame and rapidly became colonised by saltmarsh species. The site was used to demonstrate the applicability of the technique and gain understanding of the design and construction issues.
Orplands, Blackwater Estuary, Essex (Environment Agency)	1993 40 ha	The small size of Northey Island raised doubts about the applicability of lessons to larger sites, so a bigger site was progressed. Following hydrodynamic and morphological assessment, a breach retreat with engineered creeks and sacrificial wave breaks (constructed from material won from the breach and creeks) was undertaken. The site was relatively high in the tidal frame and became widely colonised by a diverse range of saltmarsh species within five years. A detailed monitoring programme was undertaken (continuing).
Pawlett Hams, Parrett Estuary, Somerset (Environment Agency)	1994 5 ha	A breach retreat enabled the site to become colonised rapidly by saltmarsh species. Bank realignment had happened historically in the area and gave confidence in adopting this approach.
Tollesbury, Blackwater Estuary, Essex (Defra, English Nature and Environment Agency)	1995 21 ha	A breach retreat on to a wheat field to be used as a demonstration site with associated R&D projects. Some wheat stubble was left to provide a degree of surface roughness and experiments were made with different surface treatments. Existing hedgerows and trees were left *in situ* to act as baffles. This was a low-elevation site, and immediately following breaching the site reverted to mudflat. Although saltmarsh establishment was initially limited, it has progressed rapidly as sedimentation has occurred. Eight years after breaching natural creeks with levees are beginning to form.
Abbotts Hall, Blackwater Estuary, Essex (visitor centre for project consortium)	1995 and 2002 20 ha and 115 ha	This site began as a tidal exchange system in 1995. Controlled sluices were inserted to a flood embankment to allow tidal inundation of the 20 ha site to increase land levels. A creek system was dug into the site to allow an outdoor testing environment with controllable water levels. Sedimentation and vegetation establishment started with the pipes before the wall was breached in 2002. The site has been developed as a larger (115 ha) demonstration area for multiple objectives including flood defence. Different land management techniques have been applied and there is a range of possible uses for the realignment sites, including alternative agriculture. Considerable design and monitoring work has been undertaken.
Blaxton Meadow, Saltram, Somerset (Environment Agency)	1995	A spillway allowed high tides to flood the site and enabled saltmarsh communities to develop. The spillway requires a continuing commitment to maintenance, and the structures have frequently been vandalised. Saltmarsh colonisation has occurred despite there being very little other saltmarsh in surrounding areas.
Thornham Bay, Chichester Harbour, Sussex (Chichester Harbour Conservancy)	1996 7 ha	A breach retreat on to a former waste tip, which was cleaned and smoothed. Creeks were excavated and a bridge constructed over the breach to maintain footpath access. The remaining parts of the breached seawall have to be maintained for health and safety reasons because of footpath access.

Site, location and county, and starting information source	Date of scheme and retreat area	Initial comments and findings
Havergate Island, River Alde, Suffolk (National Trust and RSPB)	2000 10 ha	Deteriorating walls led to a managed breach to allow saline flooding of a rough, brackish, grassland area. The site is close to bird roosting sites and has provided mudflat and saltmarsh habitat for breeding and feeding.
Black Devon Wetland, adjacent to River Forth and River Black Devon, Scotland (SEPA)	2000 7 ha	This site is based around a controlled breach of the flood embankments of the River Black Devon, to allow reclaimed saltmarsh to be re-inundated. The scheme was the first managed realignment project to take place in Scotland and was combined with ground modelling to create permanent lagoons. To date, the project is realising its aim of increasing the biodiversity of the area.
Brandy Hole, River Crouch, Essex (Environment Agency)	2002 7 ha	Several small breaches were made through an existing flood embankment, tying in with creek locations across the existing intertidal zone. A new cross-bank was required at one side of the site, but elsewhere flooding is limited by rising topography, enabling a transition of habitats to evolve. Timber footbridges were placed across each of the breaches to enable continued access along the flood embankment.
Freiston Shore, The Wash, Lincolnshire (Environment Agency, RSPB)	2002 78 ha	Realignment of a flood defence to new banks at the back of the site. The old, sub-standard defence had been built to reclaim an area of saltmarsh for agricultural use in the 1980s. Since then it had had been under severe erosion pressure and was breached in three places for flood defence reasons. The site has experienced some unpredicted changes to the fronting intertidal mudflats as the system readjusts to its previous state. There has been headward recession of new creek systems towards the breaches from the intertidal flats. Annual species have covered the site within 12 months of breaching, and significant accretion has occurred within the site. A monitoring programme is under way. A footpath was re-routed and the site is now an RSPB reserve.
Paull Holme Strays (formerly known as Thorngumbold), River Humber, Yorkshire (Environment Agency)	2003 80 ha	Realignment of a sub-standard flood defence to new banks at the back of the site. The new area of habitat provides compensation under the Habitats Regulations for losses associated with other flood defence projects locally, but will also compensate for future coastal squeeze changes. It has had to accommodate the existence of a high-pressure gas pipe. The performance of the site is not yet known. Breaching took place in August 2003 and the site has experienced a similar headward erosion of new creeks into the breaches as at Freiston Shore. This forms part of a strategic, estuary-wide, approach for the Humber for a range of purposes including flood management and conservation.

The managed realignments that have been undertaken have, on the whole, been documented. Such documents provide information on the purpose of the realignment, the design processes used and monitoring of the sites. The information has been used in the development of this guidance, but it should be used with caution becauseit has often been generated for other reporting needs. One existing document (Burd, 1995), for example, uses empirical evidence derived from the south-east of England and states that the onset of saltmarsh colonisation occurs at topographic levels of 2.1 m ODN. This absolute value is not valid in all parts of the country, since the critical determining factor in saltmarsh colonisation is topographic level relative to tidal level (ie it is

APPENDICES

dependent on frequency and duration of tidal inundation, not the absolute topographic level alone). It is clear this absolute value has been erroneously applied to some sites.

The focus of many previous schemes has been on the re-creation of saltmarsh, partly because of its inherent properties for both wave attenuation and pollutant entrapment, and also to help replace the areas of this habitat lost across the UK. Consequently, breach retreat is often the preferred implementation choice (see Section 2.8.1). It should be remembered, however, that some schemes (eg compensatory schemes for habitat change elsewhere) may require the re-creation of other habitat, such as mudflat. In these circumstances, other implementation techniques may be more appropriate. It is important to understand the project drivers to ensure the information is of relevance to a particular project.

Other techniques, such as reverse sluices or artificial or temporary spillways, have been used less frequently, but are effective for controlling inundation frequencies and duration. Vandalism has occurred on two spillway and sluice managed realignment schemes. Elsewhere active vandalism has taken the form of the removal of wire from fence lines and damage to, or removal of, warning signs. Vandalism does not necessarily reflect ease of access to the site but is an issue that needs to be allowed for in the design. If the realignment will fail or if it is likely to increase risks elsewhere as a result of vandalism, then a different design may be needed.

It is important to be aware that these are not "zero cost" schemes and can sometimes involve reasonably substantial expense on footpath diversions or environmental enhancement, in addition to scheme design, assessment and site works.

Artificial creek systems

The excavation of an artificial creek system is widely considered to be advantageous. Such engineering works can serve two purposes: (i) to enable water and its suspended sediment and propagules to flow far and wide across the site, facilitating both sedimentation and vegetation establishment; and (ii) to aid in the drainage of water from the site on the ebbing tide, thereby preventing the ponding and stagnation of water. The design of these systems may be informed through the inspection of aerial photographs to identify relict creek patterns. Material excavated from the creation of creeks (and indeed from the creation of any breaches that may be made in existing walls) can be placed elsewhere within the site to serve two purposes: (i) to be used as temporary wave breaks in areas vulnerable to wave penetration; and/or (ii) to be "sacrificial" stores of sediment that can be supplied to the managed realignment site.

Creek creation should take account of the possibility of items of archaeological interest existing along creek margins. This assumption has been applied in guidance notes, but it may not be relevant if there is no evidence of boat use, or where the creeks are too recent to have had ephemeral (nomadic) settlement in historic times.

Wind-wave action

Internally generated wind-wave action has sometimes been dismissed as being unimportant, due to the limited fetch within a breach realignment scheme. However, post-scheme monitoring tends to suggest that even these low levels of wave action can hinder sedimentation. This is an entirely natural process, but it can be a disadvantage where the aim is to promote sedimentation. Prediction of sedimentation rates on managed realignment sites is limited to numerical/computational techniques. Sediment information from schemes that have been or are being monitored will enhance modelling predictions as will measurements in the existing natural environment.

Contractors

Earth-moving contractors have experience in the techniques to implement breach realignment schemes where large volumes of material need moving. Other general contractors may need more guidance to avoid environmental damage (such as machinery getting flooded by the sea and/or releasing pollutants).

Estuaries with restricted mouths

Estuaries that have restricted mouths will respond to managed realignment through increased tidal prism. This response will also occur naturally under sea level rise. The changes at the mouth of a naturally restricted system might cause a change across a critical threshold. This possibility must be understood and evaluated to determine the consequences of the scheme.

Relocation of wildlife

High costs have been incurred in relocating water voles. Relocation may not always be necessary, however, as evidence suggests they will use realigned sites. The need should be assessed by the relevant countryside conservation agency, but it is important to appreciate that measures undertaken to date may not, necessarily, apply to future schemes as lessons are learnt and knowledge builds up.

A4.1.2   **Managed realignment schemes that have not been implemented**

It is useful to examine schemes that have been designed but not implemented. Some of the reasons that schemes have not been implemented are set out below.

1   Success criteria have been tied to specific species that cannot be guaranteed by the design.

2   Local opposition, as the public perceive that environmental concerns are being placed before people (eg direct loss of assets as a result of realignment).

3   Funding – ie too expensive to meet the success criteria. This can be a difficulty if different works are required across the site to achieve disparate success criteria and there is only one funder. Project instigators should define the money available and seek to apportion costs in respect of scheme development to different beneficiaries.

4   Existing land use has prevented progression of the scheme. Examples include a refuse landfill site inside the wall, the site being crossed by a power line and pylons, power line access, arable flagship farm, access roads and footpath, freshwater SPA and military installations.

5   Public footpaths. People want to be able walk beside the sea and this should be addressed early in the scheme. In some cases, it may be appropriate to commit to creating a new coastal footpath as part of a managed realignment scheme along or above the mean high water mark.

6   The presence of protected species such as great crested newts, water voles, reptiles, red data book species etc can cause delay, although it will probably not stop a scheme altogether.

APPENDICES

CASE STUDY EXPERIENCE IN ASSESSING PHYSICAL PROCESSES

This section demonstrates techniques that have been used on schemes to assess physical processes. This information is provided both to assist in scheme design and also to assess the performance and wider impacts of a selection of managed realignment schemes around the UK. While based upon real case studies, the information presented has been generalised so as to focus attention on techniques rather than site-specific details. Some sites have been implemented, whereas others are in progress.

Case Study 1 – Managed realignment of flood embankment within an estuary

Table A4.2   *Case study 1: summary scheme details*

Aim	Creation of mudflat habitat to compensate for change elsewhere within an estuary with Internationally designated habitats.
Site area	Approximately 40 ha (representing < 0.1 per cent of total estuary area).
Site prism	Approximately 3×10^5 m³ (representing < 0.02 per cent of total estuary tidal prism).
Location	Outer estuary
Estuary details	Large macro-tidal estuary subject to considerable previous research with good information and data availability. The intertidal area of the estuary is around 45 000 ha. The estuary has a tidal prism approaching 2×10^9 m³ and a mean freshwater input of around 200 m³/s.
Realignment method	Bank retreat
Present status	To be implemented, subject to necessary approvals.

Assessment of physical process and morphology in influencing design

The elevations of the present-day land levels within the scheme were obtained from LiDAR data and ground-truthed with a topographic survey. These elevations were used to predict the extent of habitat types within the scheme based on known elevation preferences for intertidal habitats. A GIS was used to undertake these predictions.

To create the maximum amount of sustainable mudflat habitat, the scheme design required the consideration of elevation and wave energy within the scheme.

A degree of reprofiling was recommended within the site to increase the area of the scheme below MHWN – the lower limit for marsh development. The profile was designed to provide a gentle slope from the existing mudflats to the rear of the site, thereby assisting drainage. Additionally, the scheme design opted for removal of the existing sea defences to raise wave energy within the site and limit the development of marsh vegetation. Ideally, the scheme would have removed the existing marsh in front of the seawall, to increase wave energy further. This was not possible because the marsh was designated. Breaches were therefore designed through the fronting marsh, based on a consideration of the tidal prism within the site, discharge and critical shear stress. These breaches were designed so that the velocities through them would be below those required for the erosion of muddy sediment.

As the rear of the site comprises an expanse of low-lying land, a new sea defence was required to limit the lateral extent of tidal inundation. This defence was constructed from material excavated during the reprofiling of site topography, which obviated the need to import or export material to or from the site. The design of these sea defences involved the assessment of future sea level and wave conditions. Although the scheme design was intended to create a narrow strip of marsh in front of the new sea defences, the uncertainty in the development of this marsh, plus the limited degree of wave attenuation that would be provided and uncertainty regarding the permanence of the feature, meant that it was not possible to reduce the design wave heights for the new sea defence. This will be reviewed for continuing maintenance operations.

Future evolution of the scheme

The likely future evolution of the scheme was assessed using professional expert judgement informed by a variety of existing information:

- sedimentation rates predicted from a morphodynamic numerical model
- sedimentation rates predicted at other managed realignment schemes
- history of reclamation and accretion at the site
- history of accretion of surrounding marshes
- predicted rates of sea level rise and estuary-wide response.

Assessment of design impacts on physical processes and morphology

At an early stage of the scheme, one of the statutory consultees requested that a numerical model should be used to assess the impacts of the scheme on the physical process of the estuary. As a result, detailed numerical modelling was undertaken. In addition, a desk-based assessment of changes in empirical relationships within the estuary was also undertaken.

Numerical modelling approach

The modelling approach involved characterising the existing baseline conditions within the estuary, and then simulating the conditions envisaged under the "with scheme" scenario. Comparison of the differences in modelling results between these scenarios provided an indication of the magnitude of the potential changes due to the scheme.

The numerical modelling suite used in the investigation incorporated hydrodynamics, wave and morphodynamic modelling. This suite used a curvilinear grid configuration and represented the three-dimensional nature of the estuary, with the water column represented by eight vertical layers. The model grid was refined locally in the vicinity of the proposed managed realignment scheme to provide a sufficient level of detail to enable appropriate and accurate assessments of changes in physical processes and morphology in the immediate vicinity of the site. The model was run for 15 days to incorporate a short model warm-up time and then fully capture a spring-neap tidal cycle. The output was influenced by a scaling factor, which resulted in simulation of estuary changes over a six-month period.

Output from the numerical model was interrogated and interpreted using a number of techniques. Results indicated that the effect of the scheme on existing (baseline) water levels, tidal flow vectors and sediment erosion/deposition patterns was extremely localised and relatively small in magnitude. It was concluded that the effects of the scheme on the existing physical processes and morphology of the estuary would not be significant. Output from the numerical model was also used to inform assessments of changes in slack duration and tidal excursion of the estuary. These changes were found to be insignificantly small.

Assessment of empirical relationships

Consideration was made of the changes that the proposed scheme would cause to several existing empirical relationships between various morphological and hydrodynamic parameters widely used in estuary studies. Results indicated that the changes to existing estuary "stability" (defined by the O'Brien relationship, see Section 5.3.2) and tidal asymmetry (defined by independent Dronkers and Renger relationships) were insignificantly small.

Appropriateness of techniques used and lessons learned

No specific information is provided on this point.

Scheme design

The main difficulty in assessing the physical processes for this site was the prediction of future sedimentation rates. Although various techniques were used, the long-term evolution of the scheme over 50 years is still subject to some uncertainty and has to be accounted for in the assessment of risk.

Predicted scheme impacts

Given the scale of the scheme in relation to its location and the size of the estuary, it could be argued that the assessment of empirical relationships, including some average velocity calculations through various estuary cross-sections would have been sufficient. If these calculations demonstrated significant increases in average velocity, then further modelling work could have been undertaken. The requirement for modelling reflects the primacy of stakeholder requirements in this scheme. Although numerical models have advantages, they are highly dependent upon sufficient quality and quantity of appropriate data for their calibration and reliable use, so they may not provide more accurate solutions if used in isolation. In this example, accurate topography for the proposed managed realignment site was obtained from a combination of LiDAR and topographic surveying. The bathymetry of the whole estuary was obtained from regular annual surveys, supplemented in parts by Admiralty data and additional hydrographic survey data. The Environment Agency provided water level information for calibration from around 20 sites throughout the estuary. Velocity date for model calibration was obtained from an ADCP survey plus previous survey data. Environment Agency water quality monitoring was the source of salinity data for calibration for five sites. The morphodynamic model required extensive calibration in the form of 20 years of historical chart data.

A4.2.2

Case Study 2 – Managed realignment of a gravel embankment along the open coast

Table A4.3   *Case study 2: summary scheme details*

Aim	Create a more sustainable shoreline alignment to reduce the risk of breaching during storm events.
Site details	Flat, low-lying coastal plain protected from inundation by the present, heavily modified, gravel bank.
Location	Open coast.
Realignment method	Bank retreat by 50 m landwards along a length of approximately 1–2 km.
Present status	Alternative set-back distances over greater realignment lengths are under further consideration.

Assessment of physical process and morphology in influencing design

The existing gravel bank has been extremely heavily modified through regular reprofiling activities and occasional replenishment. Its morphology and sediment composition cannot therefore be considered to be in a natural condition. Consequently, percolation, overwashing and roll-back are replaced by wave reflection of a densely compacted seaward face and erosion of bank material during storm events. Sediment supply is limited and outstripped by sediment transport, leading to swash alignment and cannibalisation of sediment in places.

The gravel bank sits on top of a sandy basement, but in the realigned position the existing backshore foundation is composed of soft, unconsolidated material. The potential exists for the realigned bank to sink into the basement, unless supported (eg by a geotextile). The precise position of the realigned bank is constrained by the presence of a stream, which, if modified, could increase the landward flood risk and is designated as a SSSI. Alternatively, this stream would also need to be relocated landwards, adding to project costs.

To manage future stability of the realigned bank, various design options were considered, including: realignment with no other intervention; realignment with timber groynes relatively closely spaced along the frontage; realigned with rock groynes at a small number of defined locations. The preferred option was selected following the numerical modelling approaches described below.

Assessment of design impacts on physical processes and morphology

The key areas for investigation were: (i) the planform stability and future evolution of the realigned shoreline; and (ii) the potential impact of the realignment on the existing longshore drift rates along the frontage.

Numerical modelling approach

The modelling approach involved consideration of shoreline plan shape evolution and longshore drift potential. The existing baseline conditions along the coastline were first characterised, and then conditions envisaged under a range of "with scheme" scenarios were simulated. Comparison of the differences in modelling results between these scenarios provided an indication of the magnitude of the potential changes due to the scheme.

The numerical modelling suite was driven by input wave and water level conditions. Unfortunately, at this site (telemetered) water level data was of a poor quality and unsuitable for use in the model. Consequently, astronomic tidal data were generated through harmonic analysis and used in the subsequent modelling. The limitation of this approach was that extreme water levels (astronomical tide plus meteorologically induced surge) were not considered in the model.

Appropriateness of techniques used and lessons learned

Using a modelling approach allowed a flexible information source to be set up to test different design scenarios and to help identify limitations in the strategic design envisaged for the site. The risks involved in failure of the shingle ridge meant this was appropriate from the outset, other methods having been employed in the initial strategy studies.

Scheme design

One of the initial observations of the consultants involved in the scheme was the limited extent of the realignment, in terms of both the set-back distance and the lateral extent of frontage along which set-back would occur. It was considered that the initial approach was driven by non-technical factors, including ensuring compliance with previous strategic study recommendations, implementing a mid-term solution rather than a long-term sustainable solution, and realignment over as small an area of land as possible. Based upon improved understanding of the physical processes and engineering design issues (technical, environmental and economic), alternative realignment distances, over different shoreline lengths, are being considered.

Scheme impacts

Numerical modelling was undertaken early in the study to provide information to aid iteration of the scheme design (ie identifying the best scheme option, based on minimal impact on existing coastal processes). This has proven useful, since further consideration is being given to alternative scheme designs and the strategy underlying this.

Although numerical models have advantages, they are highly dependent upon sufficient quality and quantity of appropriate data for their calibration and reliable use. In this instance, sufficient information was derived to provide sound advice on the approach to realignment.

A4.2.3

Case Study 3 – Managed realignment of a flood embankment within a large tidal embayment

Table A4.4   *Case study 3: summary scheme details*

Aim	Improving flood defence performance.
Site area	Approximately 80 ha (representing < 0.2 per cent of total embayment area).
Site prism	Approximately $6-7 \times 10^5$ m³.
Location	Towards landward end of tidal embayment.
Realignment method	Improvements made to secondary defences. Construction of a new cross-bank. Three 50 m-wide breaches cut into the existing outer sea defence.
Present status	Implemented.

Assessment of physical process and morphology in influencing design

The tidal prism of the site was not defined based upon a digital ground model and volumetric calculations below a defined horizontal plane (representing a given water level, eg MHWS). Instead, an alternative approach was used based upon a calculation of site area multiplied by the difference between an "average" site level and MHWS. This prism value was initially used in an empirical equation derived by Burd (1995) to calculate a breach width of the order of 110 m (assuming a land level of 2.75 m ODN).

In addition, numerical model runs were undertaken to simulate in more detail the effects of a number of breaches with widths ranging from 25 m to 50 m. A total of 18 breach width and location scenarios were investigated in these model runs. These ranged from one, two or three breaches of 25 m width, through to one, two or three breaches of 50 m width. The breach locations were related to existing creek networks across the intertidal zone. The scheme design (three 50 m breach widths) was optimised based upon model results.

Assessment of design impacts on physical processes

Due to the size of the scheme and the existing nature conservation importance of the tidal embayment adjacent to the managed realignment site, a detailed geomorphological assessment and numerical modelling exercise was undertaken to assess the potential impacts of the scheme on the wider environment.

Numerical modelling approach

Hydrodynamic, sediment and wave modelling techniques were applied. These included regional and local wave transformation modelling, assessment of locally generated wind-waves, 2D flow and sediment modelling. Results indicated that waves from deep water would not penetrate the managed realignment site under normal tidal conditions and that locally generated waves within the site would be minimal because of the limited fetch. Hydrodynamic modelling determined the number of inundations that the site would receive and identified that existing creek networks would continue to evolve. Sedimentation rates were determined based on an assessment of accretion of the existing intertidal zone and comparison with observed rates on managed realignment sites within previously implemented schemes in similar environments.

Scheme design

The modelling investigations considered a variety of breach widths and locations. The maximum considered was three breaches, each of 50 m width. Bank realignment was not considered.

Appropriateness of techniques used and lessons learned

Following scheme implementation, the engineers have witnessed unexpected impacts on the existing intertidal area. Intertidal creeks have developed from the seaward extent of the intertidal flats and carved their way back to the high water level, ie headward recession. The new creeks have receded to the breaches, which have scoured down. The material being eroded from the channels deposits along the edges of the channel and has led to accretion of approximately 1 m levees before stabilising. The cause is believed to be the lengthy period that the flow ebbs out of the site on high spring tides. On high spring tides the creeks run for most of the ebb, rather than a few hours after high tide. This suggests there may have been some limitation in the assessment techniques employed. Factors such as topography may preferentially influence the discharge through certain breaches and the discharge at lower tide levels, although small, may be sufficient to move unconsolidated mud. Not all of the subsequent effects can be attributed to uncertainties in the modelling, as this did not focus on the existing intertidal area.

Lessons learned are that, despite a full consultation and investigation, not all interests affected by a proposed managed realignment may be identified at the pre-scheme stage. In addition, more attention should be given to the effects that the managed realignment may have on the regime within the existing intertidal area. In this case, the intertidal changes affected a local shellfishery through erosion, directly damaging laid beds. These effects are some kilometres away, where the intertidal has a shallow slope and is formed of unconsolidated sediments.

Scheme impacts

The scheme has been fully achieved its primary objective of providing new and effective flood defence. In addition, it has enhanced the SAC/SPA/Ramsar site by linking a further 80 ha of habitat into the system. The site is RSPB's most successful new reserve, attracting tens of thousands of visitors in its first year and having received European funding towards the cost. The only observed detrimental effect of the scheme is on a fisheries interest, which was not identified pre-scheme and is being addressed with the local fishermen.

References

ABP MARINE ENVIRONMENTAL RESEARCH (2002). "Habitat quality measures and monitoring protocols". Proposal for Defra, ABPmer, Southampton

ABP MARINE ENVIRONMENTAL RESEARCH (2003a). *Estuaries database 2003: A spatio-temporal GIS database of environmental data for 6 UK estuaries* (CD). Defra R&D Project FD 2110. ABPmer, Southampton, for Defra and the Environment Agency

ABP MARINE ENVIRONMENTAL RESEARCH (2003b). *Modelling of a proposed managed realignment scheme at Welwick, Humber Estuary.* Report R976, ABPmer, Southampton

ABP RESEARCH & CONSULTANCY (1997a). *Design scheme for habitat creation.* Report R584(a), ABP, Southampton

ABP RESEARCH & CONSULTANCY (1997b). *Investigation into the breaching/removal of the flood embankment in the lee of Orford Ness.* Report R575, ABP, Southampton

ABP RESEARCH & CONSULTANCY (1997c). *Lantern Marshes, Orford Ness: evaluation of options for managed retreat.* Report R727, ABP, Southampton, for the National Trust

ABP RESEARCH & CONSULTANCY (1998). *Review of coastal habitat creation, restoration and recharge schemes.* Report R909, ABP, Southampton

ABP RESEARCH & CONSULTANCY (1999). *A modelling investigation into an area of managed retreat on Havergate Island, Suffolk.* Report R778a, ABP, Southampton, for Royal Society for the Protection of Birds

ADAM, P (1990). *Saltmarsh ecology.* Cambridge University Press, Cambridge

ALLEN, J R L and PYE, K (1992). *Saltmarshes: morphodynamics, conservation and engineering significance.* Cambridge University Press, Cambridge

ALLEN, J R L and RAE, J E (1988). "Vertical salt-marsh accretion since the Roman period in the Severn Estuary, southwest Britain". *Marine geology*, vol 83, pp 225–235

AMOS, L and MOSHER, D C (1985). "Erosion and deposition of fine-grained sediments from the Bay of Fundy". *Sedimentology*, vol 32, pp 815–832

ANISFELD, S C, TOBIN, M J and BENOIT G (1999). "Sedimentation rates in flow-restricted and restored salt marshes in Long Island Sound". *Estuaries*, vol 22, no 2A, pp 231–244

ASANO, T, DEGUCHI, H and KOBAYASHI, N (1992). "Interaction between water waves and vegetation". In: B L Edge (ed), *Coastal engineering 1992. Proc 23rd int conf coastal engineering, Venice, 4–9 Oct 1992.* ASCE, Reston, VA, pp 2710–2723

ASH, J R V and FENN, T (1997). "Tollesbury managed setback experimental site". In: *Seminar on managed retreat in Britain, Wallingford, 13 Nov 1997.* HR Wallingford, Wallingford

ATKINSON, P W, CROOKS, S, GRANT, A and REHFISCH, M M (2001). *The success of creation and restoration schemes in producing intertidal habitat suitable for waterbirds.* Research Report 425, English Nature, Peterborough

BAMBER, R N, GILLILAND, P M and SHARDLOW, M E A (2001). *Saline lagoons: a guide to their management and creation.* Interim version, Saline Lagoon Working Group, English Nature, Peterborough

BARBER, N F (1963). "The directional resolving of an array of wave detectors". In: *Ocean wave spectra, proc conf*, Prentice-Hall, Old Tappan, pp 137–150

BELL, A (1997). "Blaxton Meadow managed retreat". In: *Seminar on managed retreat in Britain, Wallingford, 13 Nov 1997*. HR Wallingford, Wallingford

BESLEY, P (1999). *Overtopping of seawalls: design and assessment manual*. R&D Technical Report W178, HR Wallingford, Wallingford

BOORMAN, L A (2003). *Saltmarsh review. An overview of coastal saltmarshes, their dynamic and sensitivity characteristics for conservation and management*. Report 334, Joint Nature Conservation Committee, Peterborough

BOUMANS, R M J and DAY, J W Jr (1993). "High precision measurements of sediment elevation in shallow coastal areas using a sedimentation-erosion table". *Estuaries*, vol 16, no 2, pp 375–380

BRAMPTON, A H (1992). "Engineering significance of British saltmarshes". In: J R L Allen and K Pye (eds), *Saltmarshes: morphodynamics, conservation and engineering significance*, Cambridge University Press, Cambridge, pp 115–122

BRICKER-URSO, S, NIXON, S W, COCHRAN, J K, HIRSCHBERG, D J and HUNT, C (1989). "Accretion rates and sediment accumulation in Rhode Island salt marshes". *Estuaries*, vol 12, no 4, pp 300–317

BRITISH STANDARDS INSTITUTION (2000). *Maritime structures – Part 1: code of practice for general criteria*. BS 6349:2000, BSI, London

BROOKE, J, LANDIN, M, MEAKINS, N and ADNITT, C (2000). *The restoration of vegetation on saltmarshes*. R&D Technical Report W208, Environment Agency, Bristol

BROOKS, A and AGATE, E (2000). *Sand dunes: a practical handbook*. 4th edn, British Trust for Conservation Volunteers, Wallingford

BROWN, J E, McDONALD, P, PARKER, A and RAE, J E (1999). "The vertical distribution of radionuclides in a Ribble Estuary saltmarsh: transport and distribution processes". *J environmental radioactivity*, vol 43, no 3, pp 259–275

BRYANT, J C and CHABRECK, R H (1998). "Effects of impoundment on vertical accretion of coastal marsh". *Estuaries*, vol 21, no 3, pp 416–422

BURD, F H (1995). *Managed retreat: a practical guide*. English Nature, Peterborough

CAHOON, D R, FRENCH, J R, SPENCER, T, REED, D J and MÖLLER, I (2000). "Vertical accretion versus elevational adjustment in UK saltmarshes: an evaluation of alternative methodologies". In: K Pye and J R L Allen (eds), *Coastal and estuarine environments: sedimentology, geomorphology and geoarchaeology*. Geological Society of London, Special Publication 175, pp 223–238

CARPENTER, K E (1997). "Options for reinstating tidal flooding". In: *Seminar on managed retreat in Britain, Wallingford, 13 Nov 1997*. HR Wallingford, Wallingford

CARPENTER, K E and BRAMPTON, A (1996). *Maintenance and enhancement of saltmarshes*. R&D Note 473, Environment Agency, Bristol

CARR, A P and BLACKLEY, M W L (1985). "Implications of sedimentological and hydrological processes on the distribution of radionuclides: an example of a salt marsh near Ravenglass, Cumbria". *Estuarine, coastal and shelf science*, vol 22, pp 529–543

CARR, A P and BLACKLEY, M W (1986). "Seasonal changes in surface level of a saltmarsh creek". *Earth surface processes and landforms*, vol 11, pp 427–439

CARTER, R (1990). *Coastal environments: an introduction to the physical, ecological and cultural systems of coastlines*. Academic Press, New York/London

CHADWICK, A J, POPE, D J, BORGES, J and ILIC, S (1995a). "Shoreline directional wave spectra. Part I: an investigation of spectral and directional analysis techniques". *Proc Instn Civ Engrs, water and maritime engineering*, vol 112, pp 198–208

CHADWICK, A J, POPE, D J, BORGES, J and ILIC, S (1995b). "Shoreline directional wave spectra. Part II: instrumentation and field measurements". *Proc Instn Civ Engrs, water and maritime engineering*, vol 112, pp 209–214

CHANG, Y-H, SCRIMSHAW, M D, MACLEOD, C L and LESTER, J N (2001). "Flood defence in the Blackwater Estuary, Essex, UK: the impact of sedimentological and geochemical changes on salt marsh development in the Tollesbury managed realignment site". *Marine pollution bulletin*, vol 42, no 6, Jun, pp 469–480

CHAPMAN, V J and RONALDSON, J W (1958). "The mangrove and saltmarsh flats of the Auckland Isthmus". New Zealand Department of Scientific and Industrial Research, *Bulletin*, no 125, pp 1–79

CHRISTIE, M C and DYER, K R (1998). "Measurements of the turbid tidal edge over the Skeffling mudflats". In: K S Black, D M Paterson and A Cramp (eds), *Sedimentary processes in the intertidal zone*. Geological Society of London, Special Publication 139, pp 45–55

CIRIA and CUR (1991). *Manual on the use of rock in coastal and shoreline engineering*. Special Publication 83, CIRIA, London

CLARKE, R T, GRAY, A J, WARMAN, E A and MOY, I L (1993). *Niche modelling of saltmarsh plant species*. T/04/00194/REP, Institute of Terrestrial Ecology, Huntingdon, for Energy Technology Support Unit

COASTAL GEOMORPHOLGICAL PARTNERSHIP (2001). *Coastal data analysis: The Wash. Study 2: wave attenuation over inter-tidal surfaces*. Report STCG/2000/48, for the Environment Agency, Anglian Region

COLLINS, M B, AMOS, C L and EVANS, G (1981). "Observations of some sediment transport processes over intertidal flats, the Wash, UK". In: *Proc Int Assoc Sedimentologists Congress, Texel*. IAS Special Publication 5, pp 81–98

COOPER, N J (2003). "The use of 'managed retreat' in coastal engineering". *Proc Inst Civ Engrs, Engineering sustainability*, vol 1, no 2

COOPER, N J (in press). "Wave dissipation across inter-tidal surfaces in the Wash tidal inlet, eastern England". *J coastal research*

COVENTRY, S and WOOLVERIDGE, C (1999). *Environmental good practice on site*. C502, CIRIA, London

CROOKS, S (1999). "A mechanism for the formation of overconsolidated horizons within estuarine floodplain alluvium: implications for the interpretation of Holocene sea-level curves". In: J Alexander, S B Marriott and R Hey (eds), *Floodplains: interdisciplinary approaches*, Geological Society of London, Special Publication 163, pp 197–215

DAVIS, LANGDON and EVEREST (2002). *Spon's civil engineering and highway works price book*. 16th edn, Spon Press, London

DEAN, R G (1971). *Hydraulics of inlets*. COEL/UFL-71/019, Department of Coastal and Oceanographic Engineering, University of Florida, Gainsville

DEAN, R G (1979). "Effects of vegetation on shoreline erosional processes". In: P E Gregson, J R Clark and J E Clark (eds), *Wetland functions and values: the state of our understanding. Proc nat symp on wetlands*, American Water Association, Minneapolis, pp 415–426

DEFRA (1999). *High level targets for flood and coastal defence and elaboration of the Environment Agency's flood defence supervisory duty*. Department for Environment, Food and Rural Affairs, London. <http://www.defra.gov.uk/environ/fcd/hltarget/default.htm>

DEFRA (2000, rev 2003). *Flood and coastal defence project appraisal guidance: economic appraisal*. FCDPAG3, Department for Environment, Food and Rural Affairs, London <http://www.defra.gov.uk/environ/fcd/pubs/pagn/fcdpag3/default.htm>

DEFRA (2001). *Shoreline management plans: a guide for coastal defence authorities*. Department for Environment, Food and Rural Affairs, London. <http://www.defra.gov.uk/environ/fcd/pubs/smp/revisedsmpguidancefinal.pdf>

DEFRA (2002). *Flood Management Division scheme prioritisation system*. Department for Environment, Food and Rural Affairs, London. <http://www.defra.gov.uk/corporate/regulat/forms/flood/LDW14.pdf>

DEFRA (2003a). *Managed realignment: land purchase, compensation and payment for alternative beneficial land use*. Department for Environment, Food and Rural Affairs, London. <http://www.defra.gov.uk /environ/fcd/policy/managed realignmentCOMP4.htm>, accessed April 2003

DEFRA (2003b). *Procedural guidance for production of shoreline management plans*. Interim guidance, May, Department for Environment, Food and Rural Affairs, London. <http://www.defra.gov.uk/corporate/consult/smpguidance/consultdoc.pdf>

DELO, E A (1988). *Estuarine muds manual*. Report SR 164, Hydraulics Research Laboratory, Wallingford

DeVRIES, J W (1992). "Field measurements of the erosion of cohesive sediments". *J coastal research*, vol 8, no 2, pp 312–318

DIXON, A M and WEIGHT, R C (1997). "Managing coastal realignment: case study at Orplands sea wall, Blackwater Estuary, Essex". In: *Seminar on managed retreat in Britain, Wallingford, 13 Nov 1997*. HR Wallingford, Wallingford

DIXON, M J and TAWN, J A (1997). "Spatial analyses for the UK coast". Internal document 112.04, Proudman Oceanographic Laboratory, Birkenhead

DIXON, A M, LEGGETT, D J and WEIGHT, R C (1998). "Habitat creation opportunities for landward coastal realignment: Essex case studies". *J CIWEM*, vol 12, no 2, pp 107–112

DOODY, J P (2001). *Coastal conservation and management: an ecological perspective*. Conservation Biology series no 13, Kluwer Academic Publishers, Boston, 306 pp

DOODY, J P and RANDALL, R E (2003). *A guide to the management and restoration of coastal vegetated shingle*. English Nature, Peterborough

DRONKERS, J (1998). "Morphodynamics of the Dutch delta". In: J Dronkers and M Scheffers (eds), *Physics of estuaries and coastal seas. Proc 8th int biennial conf estuaries and coastal seas, The Hague, 9–12 Sep 1996*. AA Balkema, Rotterdam, pp 297–304

DYER, K R (1979). *Estuarine hydrography and sedimentation: a handbook*. Cambridge University Press, Cambridge

DYER, K R (1986). *Coastal and estuarine sediment dynamics*. John Wiley & Sons, Chichester

EMPHASYS CONSORTIUM (2000a). *Modelling estuary morphology and process. Final report*. Report TR 111, HR Wallingford, Wallingford, for MAFF

EMPHASYS CONSORTIUM (2000b). *A guide to prediction of morphological change within estuarine systems. Final report*. Report TR 114, HR Wallingford, Wallingford, for MAFF

ENGLISH NATURE (1992–96). *Northey Island managed retreat*. Research Reports 6 (1992), 61 (1993), 103 (1994), 157 (1995) and 185 (1996); Project Reports 2, 3, 4, 5 and 6. English Nature, Peterborough

ENGLISH NATURE (2004). *England's best wildlife and geological sites – the condition of Sites of Special Scientific Interest in England in 2003*. ST10.4, English Nature, Peterborough

ENGLISH NATURE, ENVIRONMENT AGENCY, DEFRA, LIFE and NERC (2003). *Living with the sea. Coastal habitat restoration: towards good practice*. English Nature, Peterborough. <http://www.english-nature.org.uk/livingwiththesea/project_details/good_practice_guide/Home.htm>

ENVIRONMENT AGENCY, ENGLISH NATURE, RSPB and HM PRISON SERVICE (2004). *The Wash Banks*. Scheme leaflet BGXU, Environment Agency, Peterborough

ENVIRONMENTAL CONSULTANCY UNIVERSITY OF SHEFFIELD (ECUS) (2003). *Guidance for coastal defence design in relation to their landscape and visual impacts (final report)*. Contract Science Report 531, ECUS, Sheffield, for Countryside Council for Wales

ESCOFFIER, F F (1940). "The stability of tidal inlets". *Shore and beach*, vol 8, no 4, pp 114–115

FONSECA, M S and CAHALAN, J A (1992). "A preliminary evaluation of wave dissipation by four species of seagrass". *Estuarine, coastal and shelf science*, vol 35, pp 565–576

FRENCH, C E, FRENCH, J R, CLIFFORD, N J and WATSON, C J (2000). "Sedimentation-erosion dynamics of abandoned reclamations: the role of waves and tides". *Continental shelf research*, vol 20, nos 12–13, pp 1711–1733

FRENCH, J R (1995). "Function and optimal design of saltmarsh channel networks". In: *Saltmarsh management for flood defence. Proc research seminar, Nov 1995*. Project Record 480/1/SW, National Rivers Authority, pp 85–95

FRENCH, J R (1996). "Function and design of tidal channel networks in restored saltmarshes". In: P Gardiner (ed) *Proc conf Tidal '96, University of Brighton, 12–13 Nov*, pp 128–137

FRENCH, J R and BURNINGHAM, H (2003). "Tidal marsh sedimentation versus sea-level rise: a southeast England estuarine perspective". In: *Proc coastal sediments '03, 18–23 May, Clearwater Beach, Florida*, East meets West Productions, Corpus Christi, Texas, pp 1–14

FRENCH, J R, SPENCER, T, MURRAY, A L and ARNOLD, N S (1995). "Geostatistical analysis of sediment deposition in two small tidal wetlands, Norfolk, UK". *J coastal research*, vol 11, no 2, pp 308–321

FRIEDRICHS, C T and AUBREY, D G (1988). "Non-linear tidal distortion in shallow well-mixed estuaries. A synthesis". *Estuarine, coastal and shelf science*, vol 27, no 5, pp 521–546

GAO, S and COLLINS, M B (1994). "Tidal inlet equilibrium in relation to cross-sectional area and sediment transport patterns". *Estuarine, coastal and shelf science*, vol 27, no 5, pp 521–545

GARBUTT, A, GRAY, A, READING, C, BROWN, S and WOLTERS, M (2003). "Saltmarsh and mudflat development after managed realignment". In: *Proc 38th Defra flood and coastal management conf, Jul 2003, Keele*, pp 1–10

GARBUTT, R A, READING, C J, WOLTERS, M, GRAY, A J and ROTHERY, P (2004). "Monitoring the development of intertidal habitats on former agricultural land after the managed realignment of coastal defences at Tollesbury, Essex, UK". Paper submitted to *Marine pollution bulletin*

GRAY, A J and MOGG, R J (2001). "Climate impacts on pioneer saltmarsh plants". *Climate research*, vol 18, pp 105–112

HALCROW CONSULTING ENGINEERS (1994, rev 2001). *A guide to the understanding and management of saltmarshes*. R&D Note 324, Environment Agency, Bristol

HALCROW GROUP (1999). *Wash Banks. Hobhole to Butterwick Low hydrodynamic, geomorphic and environmental assessment*. Halcrow Group, Swindon

HALCROW GROUP (2002a). *Futurecoast*. Halcrow Group, Swindon, for Defra and National Assembly of Wales. <www.defra.gov.uk/environ/fcd/research/futurecoast.htm>

HALCROW GROUP (2002b). *Implementing managed retreat as a strategic flood and coastal defence option*. R&D project FD 2008, Defra, London, and Environment Agency, Bristol

HANSEN, D V and POULAIN, P-M (1996). "Quality control and interpolations of WOCE/TOGA drifter data". *J atmospheric and oceanic technology*, vol 13, pp 900–909

HARRISON, E Z and BLOOM, A L (1977). "Sedimentation rates on tidal salt marshes in Connecticut". *J sedimentary petrology*, vol 47, pp 1484–1490

HASSELMANN, K, BARNETT, T P, BOUWS, E, CARLSON, H, CARTWRIGHT, D E, ENKE, K, EWING, J A, GIENAPP, H, HASSELMANN, D E, KRUSEMAN, P, MEERBURG, A, MÜLLER, P, OLBERS, D J, RICHTER, K, SELL, W and WALDEN, H (1973). "Measurements of wind-wave growth and swell decay during the Joint North Sea Wave Project (JONSWAP)". *Deutsches hydrographisches Zeitschrift*, vol 12 Supplement A8, 95 pp

HAUBRICH, R A (1968). "Array design". *Bulletin of the Seismological Society of America*, vol 58, no 3, pp 979–991

HAWKE, C J and JOSÉ, P V (1996). *Reedbed management for commercial and wildlife interests*. Reserve Management Information Sheet no 4, RSPB, Sandy

HOUGH, A, SPENCER, C, LOWTHER, S and MUDDIMAN, S (1999). *Definition of the extent and vertical range of saltmarsh*. R&D Technical Report TR W153, Environment Agency, Bristol

HR WALLINGFORD (forthcoming). *Reducing the risk of embankment failure under extreme conditions – good practice review*. Defra R&D project FD 2411, HR Wallingford, Wallingford, for Defra and the Environment Agency

INGHAM, A E, rev ABBOTT, V J (1992). *Hydrography for the surveyor and engineer*. 3rd edn, Blackwell Scientific Publications, Oxford

INGLIS, C C and KESTNER, F J T (1958). "Changes in The Wash as affected by training walls and land-claim works". *Proc Instn Civ Engrs*, vol 11, pp 435–466

JOINT NATURE CONSERVATION COMMITTEE (1996). *Guidelines for the selection of biological SSSIs; inter-tidal marine habitats and saline lagoons*. JNCC, Peterborough

KEARNEY, M S and WARD, L G (1986). "Accretion rates in brackish marshes of a Chesapeake Bay estuarine tributary". *Geo-Marine Letters*, vol 6, pp 41–49

KENTULA, M E (2000)."Perspectives on setting success criteria for wetland restoration". *Ecological engineering*, vol 15, nos 3–4, pp 199–209

KNUTSON, P L and ALLEN, H H (1990). *Guidelines for vegetative erosion control on wave-impacted coastal dredge material sites*. US Army Corps of Engineers

KNUTSON, P L, BROCHU, R A, SEELIG, W N and INSKEEP, M (1982). "Wave damping in *Spartina alterniflora* marshes". *Wetlands*, vol 2, pp 87–104

KOBAYASHI, N, RAICHLE, A W and ASANO, T (1993). "Wave attenuation by vegetation". *J waterway, port, coastal and ocean engineering*, vol 119, no 1, pp 30–48

KOMAR, P D (1998). *Beach processes and sedimentation*. Prentice-Hall, Englewood Cliffs

KOMAR, P D and MILLER, M C (1975). "On the comparision between the threshold of sediment motion under waves and unidirectional currents with a discussion of the practical evaluation of the threshold". *J sedimentary petrology*, vol 45, pp 362–367

LAMBERTH, C and HAYCOCK, N (2002). *Regulated tidal exchange: an intertidal habitat creation technique*. Report by Haycock Associates Ltd for RSPB and Environment Agency

LEGGETT, D J and DIXON, A M (1994). "Management of the Essex saltmarshes for flood defence". In: R Falconer and P Goodwin (eds), *Wetland management*, Institution of Civil Engineers, London, pp 232–245

LEGGETT, D J and HOLLIDAY, E M (2002). "Reducing the impact of construction in the coastal and marine environment". In: F V Gomes et al (eds), *Littoral 2002. 6th int symp proc: a multi-disciplinary symposium on coastal zone research, management and planning, Oporto, 22–26 Sep 2002*, vol II, pp 551–558

LEGGETT, D J, BUBB, J M and LESTER, J N (1995). "The role of pollutants and sedimentary processes in flood defence. A case study: salt marshes of the Essex coast, U.K.". *Environmental technology*, vol 16, pp 457–466

LETZSCH, W S and FREY, R W (1980). "Deposition and erosion in a Holocene saltmarsh, Sapelo Island, Georgia." *J sedimentary petrology*, vol 50, no 2, pp 529–542

McCONNELL, K (1998). *Revetment systems against wave attack – a design manual*. Thomas Telford, London, for HR Wallingford

MEAKINS, N C, BUBB, J M and LESTER, J N (1995). "The mobility, partitioning and degradation of atrazine and simazine in the salt marsh environment". *Marine pollution bulletin*, vol 30, no 12, pp 812–819

MILLARD, K and SAYERS, P (2000). *Maximising the use and exchange of coastal data. A guide to best practice*. C541, CIRIA, London

MILLER, M C, McCAVE, I N and KOMAR, P D (1977). "Threshold of sediment motion under unidirectional currents". *Sedimentology*, vol 24, pp 507–527

MÖLLER, I, SPENCER, T and RAWSON, J (2002). "Spatial and temporal variability of wave attenuation over a UK east-coast saltmarsh". In: *Proc 38th int conf coastal engineering, Cardiff, Jul 2002*

MÖLLER, I, SPENCER, T and FRENCH, J R (1996). "Wind wave attenuation over saltmarsh surfaces: preliminary results from Norfolk, England". *J coastal research*, vol 12, no 4, pp 1009–1016

MÖLLER, I, SPENCER, T, FRENCH, J R, LEGGETT, D J and DIXON, M (1999). "Wave transformation over salt marshes: a field and numerical modelling study from North Norfolk, England." *Estuarine, coastal and shelf science*, vol 49, no 3, pp 411–426

MÖLLER, I, SPENCER, T, FRENCH, J R, LEGGETT, D J and DIXON, M (2001) "The sea-defence value of saltmarshes – a review in the light of field evidence from North Norfolk". *J CIWEM*, vol 15, pp 109–116

NATURE CONSERVANCY COUNCIL (1989). *Guidelines for the selection of biological SSSIs*. JNCC, Peterborough

NIELSEN, N (1935). "Eine Methode zur exakten Sedimentationsmessung: Studien über die Marschbildung auf der Halbinsel Skallingen". Det Kgl Danske Videnskabernes Selskab, *Biol med*, vol 12, no 4, pp 1–97

NORDSTROM, K F, PSUTY, N P and CARTER, R W G (eds) (1990). *Coastal dunes: processes and morphology*. J Wiley & Sons, Chichester

NUTTALL, P M, BOON, A G and ROWELL, M R (1998). *Review of the design and management of constructed wetlands*. Report 180, CIRIA, London

O'BRIEN, M P (1931). "Estuary tidal prisms related to entrance areas". *Civil engineering*, vol 1, no 8, pp 738–739

O'BRIEN, M P (1966). "Equilibrium flow areas of tidal inlets on sandy coasts". In: *Proc 10th coastal engg conf, Tokyo*, ASCE, New York, vol 1, pp 676–686

OLIVER, F W (1929). Spartina problems. *Ann appl bot*, vol 7, pp 25–39

OWEN, M W (1984). "Effectiveness of saltings in coastal defence". *Proc MAFF conf of river and coastal engineers, Cranfield*

PERROW, M R and DAVY, A J (eds) (2002). *Handbook of ecological restoration. Vol 1 Principles of restoration; Vol 2 Restoration in practice*. Cambridge University Press, Cambridge

PESTRONG, R (1969). "The shear strength of tidal marsh sediments". *J sedimentary petrology*, vol 39, no 1, pp 322–394

PETHICK, J S (1984). *An introduction to coastal geomorphology*. Edward Arnold, London

PETHICK, J and BURD, F (1996). *Inter-tidal restoration: aims and techniques*. Report to ABP Research & Consultancy. Coastal Research Unit, University of Cambridge, Cambridge, pp 1–15

PETTS, J (ed) (1995). *Environmental assessment: good practice. Proc CIEF conf good practice in environmental assessment, London, 7 Sep 1995*. Special Publication 126, CIRIA, London

PIERSON, W J and MOSKOWITZ, L (1964). "A proposed spectral form for fully developed wind seas based on the similarity theory of S A Kitaigorodskii". *J geophysical research*, vol 69, no 24, pp 5181–5190

PONTEE, N I and TOWNEND, I H (1999). "The development of a cause consequence model for an estuary system". In: *Proc 34th MAFF conf river and coastal engineers, Keele University, 30 Jun–2 Jul 1999*, pp 5.2.1–5.2.17

POSFORD DUVIVIER (1997). *Wash Banks environmental statement: Gibraltar Point to Hobhole Sluice*. Final report for Environment Agency, Anglian Region. Posford Duvivier, Peterborough

POSFORD HASKONING (2002). *Trimley managed retreat site. Annual monitoring report (2000–2002), final report*. Posford Haskoning, Peterborough, for Harwich Haven Authority

RANWELL, D S (1964). "*Spartina* salt marshes in southern England. II: Rate and seasonal pattern of sediment accretion". *J ecology*, vol 52, no 1, pp 79–94

RANWELL, D S and BOAR, R (1986). *Coastal dune management guide*. Institute of Terrestrial Ecology (NERC), Huntingdon

REED, D J (1988). "Sediment dynamics and deposition in a retreating coastal salt marsh". *Estuarine, coastal and shelf science*, vol 26, pp 67–79

REED, D J, SPENCER, T, MURRAY, A L, FRENCH, J R and LEONARD, L (1999). "Marsh surface sediment deposition and the role of tidal creeks: implications for created and managed coastal marshes". *J coastal conservation*, vol 5, no 1, pp 81–90

REINECK, H E and SINGH, I B (1973). *Depositional sedimentary environments, with reference to terrigenous clastics*. 2nd edn, Springer-Verlag, Berlin

RENGER, E and PARTENSKY, H-W (1974). "Stability criteria for tidal basins". In: *Proc 14th int conf coastal engineering, Copenhagen*, American Society of Civil Engineers, Reston, pp 1605–1618

ROBERTS, W (1992). *Fluidisation of mud by waves: development of a mathematical model of fluid mud in the coastal zone.* Report SR 296, HR Wallingford, Wallingford

RODWELL, J S (ed) (2000). *British plant communities. Vol 5: Maritime communities and vegetation of open habitats.* Cambridge University Press, Cambridge, 512 pp

ROUSE, H (1939). "Discussion of 'Laboratory investigation flume traction and transportation' ". *Trans Am Soc Civ Eng*, vol 104, pp 1303–1308

RUDLAND, D J, LANCEFIELD, R M and MAYELL, P N (2001). *Contaminated land risk assessment. A guide to good practice.* Publication C552, CIRIA, London

SCHETTINI, C A F (2002). "Near bed sediment transport in the Itajaí-açu River estuary, southern Brazil". In: J C Winterwerp and C Kranenburg (eds), *Fine sediment dynamics in the marine environment.* Elsevier Science, Amsterdam, pp 499–512

SCHOOT, P M and DE JONG, J E A (1982). *Sedimentatie en erosiemetingen met behulp van de Sedi-Eros-Tafel (SET).* Note DDMI-82.401, Ministerie van verfkeer en waterstaat, Rijkswaterstaat, 7 pp

SCOTTISH NATURAL HERITAGE (2000). *Beach dunes: a guide to managing coastal erosion in beach/dune systems.* Scottish Natural Heritage, Redgorton, Perth, 128 pp

SEABERGH, W C and KRAUS, N C (1997). *PC program for coastal inlet stability analysis using Escoffier method.* Coastal and Hydraulics Engineering Technical Note ERDC/CHL CHETN IV-11. US Army Engineer Research and Development Center, Vicksburg

SELIM YALIN, M and FERREIRA DA SILVA, A M (2001). *Fluvial processes.* Monograph, International Association of Hydraulic Engineering and Research, Delft

SHORT, F T, BURDICK, D M, SHORT, C A, DAVIS, R C and MORGAN, P A (2000). "Developing success criteria for restored eelgrass, salt marsh and mud flat habitats". *Ecological engineering,* vol 15, nos 3–4, Jul, pp 239–252

SIMM, J D (ed) (1996). *Beach management manual.* Report 153, CIRIA, London

SLEATH, J F A (1984). *Sea bed mechanics.* J Wiley & Sons, New York, 334 pp

SOULSBY, R L (1998). *Dynamics of marine sands: a manual for practical applications.* Thomas Telford, London

SPENCER, T and MÖLLER I (1996). *Wave dissipation over salt marsh surfaces.* Operational Investigation Report OI/569/3/A, Environment Agency Anglian Region

SPENCER, T, MÖLLER, I and FRENCH, J R (2003). *Wave attenuation over saltmarshes.* R&D project W5B-022, Environment Agency, Bristol

STEARNS, L A and MacCREARY, D (1957). "The case of the vanishing brick dust, contribution to knowledge of marsh development". *Mosquito news,* vol 17, no 4, pp 303–304

STEERS, J A (1948). "Twelve years' measurement of accretion on Norfolk salt marshes". *Geol mag,* vol 85, pp 163–166

STEVENSON, J C, KEARNEY, M S and PENDLETON, E C (1985). "Sedimentation and erosion in a Chesapeake Bay brackish marsh system". *Marine geology,* vol 67, pp 213–235

STODDART, D R, REED, D J and FRENCH, J R (1989). "Understanding salt-marsh accretion, Scolt Head Island, Norfolk, England". *Estuaries,* vol 12, no 4, pp 228–236

STUMPF, R P (1983). "The process of sedimentation on the surface of a salt marsh". *Estuarine, coastal and shelf science*, vol 17, pp 495–508

THOMAS, K and CHESHER, T (2002). "Proposed managed re-alignment scheme Abbotts Hall, Essex: numerical modelling and project approach". In: N W H Allsop (ed), *Solving coastal conundrums. Abstracts from 28th int conf coastal engineering, Cardiff, 7–12 Jul 2002*, Thomas Telford, London

TOLHURST, T J, BLACK, K S, SHAYLER, S A, MATHER, S, BLACK, I, BAKER, K and PATERSON, D M (1999). "Measuring the *in situ* erosion shear stress of intertidal sediments with the cohesive strength meter (CSM)". *Estuarine, coastal and shelf science*, vol 49, pp 281–294

TOLHURST, T J, RIETHMÜLLER, R and PATERSON, D M (2000). "*In situ* versus laboratory analysis of sediment stability from intertidal mudflats". *Continental shelf research*, vol 20, pp 1317–1334

TOWNEND, I H (2002). "Identifying change in estuaries". In: F V Gomes et al (eds), *Littoral 2002. 6th int symp proc: a multi-disciplinary symposium on coastal zone research, management and planning, Oporto, 22–26 Sep 2002*, vol II, pp 235–243

TROW, S and MURPHY, P (2003). *Coastal defence and the historic environment. English Heritage guidance*. English Heritage, Swindon. <http://www.english-heritage.org.uk/ Filestore/policy/pdf/countryside/CoastalDefenceEH.pdf>

US ARMY CORPS OF ENGINEERS (1984). *Shore protection manual, vols I and II*. 4th edn, USACE, Washington DC

VAN DE KREEKE, J (1992). "Stability of tidal inlets: Escoffier's analysis". *Shore and beach*, vol 60, no 1, pp 9–12

VAN DER MEULEN, F (1990). "European dunes: consequences of climate change and sealevel rise". In: Th W Bakker, P D Jungerius and J A Klijn (eds), *Dunes of the European coasts*, Supplement 18, Catena Verlag, Cremlingen, pp 209–223

VAN DER MEULEN, F, JUNGERIUS, P D and VISSER, J H (eds) (1989). *Perspectives in coastal dune management*. SPB Academic Publishing, The Hague, pp 207–216

VAN DER MEULEN, F and VAN DER MAAREL, E (1989). "Coastal defence alternatives and nature development perspectives". In: F Van der Meulen, P D Jungerius and J H Visser (eds), *Perspectives in coastal dune management*, SPB Academic Publishing, The Hague, pp 183–195

VAN PROOSDIJ, D, OLLERHEAD, J and DAVIDSON-ARNOTT, R G D (2000). "Controls on suspended sediment deposition over single tidal cycles in a macrotidal saltmarsh, Bay of Fundy, Canada". In: K Pye and J R L Allen (eds), *Coastal and estuarine environments: sedimentology, geomorphology and geoarchaeology*. Geological Society of London, Special Publication 175, pp 43–57

VASSEUR, B and HÉQUETTE, A (2000). "Storm surges and erosion of coastal dunes between 1957 and 1988 near Dunkerque (France), southwestern North Sea". In: K Pye and J R L Allen (eds), *Coastal and estuarine environments: sedimentology, geomorphology and geoarchaeology*. Geological Society of London, Special Publication 175, pp 99–107

WALMSLEY, C A (2002). "Beaches" . In: M R Perrow and A J Davy (eds), *Handbook of ecological restoration*. Cambridge University Press, Cambridge, vol 2, pp 197–213

WALMSLEY, C A and DAVY, A J (2001). "Habitat creation and restoration of damaged shingle communities". In: J R Packham, R E Randall, R S K Barnes and A Neal, *Ecology and geomorphology of coastal shingle*, Westbury Academic and Scientific Publishing, Otley, pp 409–420

WAYNE, C J (1976). "The effects of sea and marsh grass on wave energy". *Coastal research notes*, vol 4, no 7, pp 6–8

WHITEHOUSE, R J S and MITCHENER, H J (1998). "Observations of the morphodynamic behaviour of an intertidal mudflat at different timescales". In: K S Black, D M Paterson and A Cramp (eds), *Sedimentary processes in the intertidal zone*. Geological Society of London, Special Publication 139, pp 255–271

WHITEHOUSE, R J S, SOULSBY, R L, ROBERTS, W and MITCHENER, H J (2000). *Dynamics of estuarine muds: a manual for practical applications*. Thomas Telford, London

WOOD, M E, KELLEY, J T and BELKNAP, D F (1989). "Patterns of sediment accumulation in the tidal marshes of Maine". *Estuaries*, vol 12, no 4, pp 237–246

WOOD, R G, BLACK, K S and JAGO, C F (1998). "Measurements and preliminary modelling of current velocity over an intertidal mudflat, Humber estuary, UK". In: K S Black, D M Paterson and A Cramp (eds), *Sedimentary processes in the intertidal zone*. Geological Society of London, Special Publication 139, pp 167–175

WOODHOUSE, W W (1982). "Coastal sand dunes of the U.S.". In: R R Lewis (ed), *Creation and restoration of coastal plant communities*. CRC Press, Boca Raton, Florida, pp 1–44

ZEDLER, J B (1984). *Salt marsh restoration: a guidebook for southern California*. Department of Biology, San Diego State University, La Jolla. California Sea Grant College, Project Report 7-CSGCP-009

ZEDLER, J B and ADAM, P (2002). "Saltmarshes". In: M R Perrow and A J Davy (eds), *Handbook of ecological restoration*. Cambridge University Press, Cambridge, vol 2, pp 238–266

Unpublished reports and projects

Note: these documents are available from the client organisation for which they were produced.

Reports for Defra

READING, C J *et al* (2001). *Managed realignment at Tollesbury and Saltram. Annual report for 2000*. Centre for Ecology and Hydrology, Monks Wood

READING, C J *et al* (2002). *Managed realignment at Tollesbury and Saltram. Final report 1995 to 2002*. Centre for Ecology and Hydrology, Monks Wood

WATTS, C W (2002). "Soil strength and stability of the managed realignment site at Tollesbury. V: Summary of measurements made during 1995/2002". In: C J Reading *et al*, *Managed realignment at Tollesbury and Saltram. Final report*. Defra-NERC contract, CSA 2313. Defra, London

Defra and Environment Agency projects

ABP MARINE ENVIRONMENTAL RESEARCH (ongoing). Habitat quality measures and monitoring protocols. R&D Project FD1918

CEFAS (ongoing). Suitability criteria for habitat creation. R&D Project FD1917

Reports for Environment Agency

PETHICK, J S (1998). Managed realignment at Thorngumbald

HALCROW GROUP (2002c). Thorngumbald Flood Alleviation Scheme. Environmental Action Plan, February 2002

HR WALLINGFORD (1999). Results of post-breach monitoring of Orplands Coastal Realignment, August 1998 to March 1999

HR WALLINGFORD (2001). *Sustainable defences. Monitoring of retreat and recharge sites.* Managed Realignment Project D 21110. Abbott's Hall, Numerical Modelling. Report EX 4367

IMPERIAL COLLEGE (1992). A survey of metal and organic micropollutant contamination in sediments of East Anglian salt marshes, 1991–1992. Report to the National Rivers Authority, Anglian Region (Eastern Area)

Report for University of Cambridge

FRENCH, J (1987). *Water and sediment fluxes in coastal salt marshes.* Research report, University of Cambridge

Report for Royal Society for the Protection of Birds

RSPB (2000). *"Seas of change": an evaluation of potential sites for intertidal habitat creation*

Websites

www[1] www.countryside.gov.uk/heritagecoasts/

This website shows the areas of heritage coastline around England and details each heritage coastline. It provides information on the Countryside Countryside Character Initiative linked to AONB

www[2] www.ccw.gov.uk

Species, habitats, earth science, landscapes etc in Wales are covered in the Countryside Council for Wales website. Its content also includes details of publications, research, news and meetings

www[3] www.english-nature.org.uk/livingwiththesea/champs

This section of the English Nature website gives information about CHaMPs under the "Living with the sea" LIFE project, including details of seven trial CHaMPs. Coastline documents about CHaMPs can be downloaded in PDF format

www[4] www.ukbap.org.uk

An online resource about biodiversity action plans, which provides information on UK biodiversity, local, species and habitat action plan summaries, and biodiversity groups and organisations. There are also downloadable reports, guidance notes, news and documents from the library

www[5] www.english-nature.org.uk/baps/intro.htm

This site giving detailed information on BAPs allows users to make specific species or habitat searches. It also covers natural area targets and species recovery programmes

www[6] www.mceu.gov.uk

Information provided by the Marine Consents and Environmental Unit includes coastal defence, dredging and disposal, energy and resources, fish and cetaceans, marine science, environment and conservation, marine construction and regulatory controls

www[7] www.environment-agency.gov.uk

The EA website contains details of regional information on the environment, business and industry, and science and research, with a link to a publications list. It also provides updates on news bulletins about the environment

www[8] www.crownestate.co.uk

This website provides information on Crown Estate news and publications

www[9] www.portoflondon.co.uk

The website provides information covering commercial, leisure and maritime uses of ports, as well as news updates from the Port of London

www[10] www.english-heritage.org.uk/

The English Heritage website includes information on; conserving historic places, archaeology, heritage, regional information and policy. There is also a link to English Heritage publications

www[11] www.English-nature.org.uk/livingwiththesea

English Nature's "Living with the sea" project website provides links and downloadable reports

www[12] www.cefasdirect.co.uk

Information about environmental activities and publications may be found on the Centre for Environment, Fisheries and Aquaculture Science (CEFAS) website. Subjects covered include contract research, consultancy, advice and training in fisheries science and management, marine environmental protection, aquaculture, and fish and shellfish disease

www[13] www.defra.gov.uk/

The Department for Environment Food and Rural Affairs (Defra) website carries information on: animal health and welfare, economics/statistics, environmental protection, exports and trade, farming, fisheries, food and drink, horticulture, plants and seeds, rural development, science, sustainable development, water, wildlife and countryside. There are links to latest news and publications. Information for flood and coastal defence can be found under www.defra.gov.uk/environ/fcd/

www[14] www.metoffice.com/construction/

The Meteorological Office can provide a construction company with a short-range weather forecast for a specific location. This can help in planning site activities and assessing the likelihood of adverse weather or other events (such as a tidal surge)

www[15] http://www.mvm.co.uk/planningofficers/shared/files/POSW

This site contains information about legislative context for planning policy to the local government and energy and utilities sector

www[16] www.bgs.ac.uk/geoindex/home.html

Geological information from a map-based index to datasets collected or obtained from a variety of sources are available from the British Geological Society website

EU legislation

Directive 79/409/EEC on the conservation of wild birds. *Official Journal of the European Commission*, L 103, 25 April 1979

Directive 79/923/EEC on the quality required of shellfish waters. *Official Journal of the European Commission*, L 281, 10 November 1979

Directive 92/43/EEC on the conservation of natural habitats and of wild flora and fauna. *Official Journal of the European Commission*, L 206, 22 July 1992

Directive 2000/60/EC establishing a framework for Community action in the field of water policy. *Official Journal of the European Commission*, L 327, 22 December 2000

UK legislation

Ancient Monuments and Archaeological Areas Act 1979 (1979 c. 46)

Coast Protection Act, 1949 (12, 13 & 14 Geo 6 c. 49)

The Coast Protection (Notices) (Wales) Regulations 2003 (Welsh SI 2003 no 1847)

The Coast Protection (Notices) (Scotland) Amendment Regulations 1996 (SI 1996 no 141)

The Conservation (Natural Habitats, &c.) Regulations 1994 (SI 1994 no 2716)

The Construction (Design and Management) Regulations 1994 (SI 1994 no 3140)

Countryside and Rights of Way Act 2000 (2000 c. 37)

Environment Act 1995 (1995 no c. 25)

The Environmental Impact Assessment (Land Drainage Improvement Works) Regulations 1999 (SI 1999 no 1783)

The Environmental Impact Assessment (Scotland) Regulations 1999 (Scottish SI 1999 no 1)

Flood Prevention and Land Drainage (Scotland) Act 1997 (1997 c. 36)

Flood Prevention (Scotland) Act, 1961

Food and Environment Protection Act 1985 (1985 c. 48)

The Harbour Works (Environmental Impact Assessment) Regulations 1999 (SI 1999 no 3445)

Highways Act 1980 (1980 c. 66)

Land Drainage Act 1991 (199 c. 59)

The Land Drainage (Grants) Regulations 1967 (SI 1967 no 212)

Merchant Shipping Act 1988 (1988 c. 12)

The Planning (Environmental Impact Assessment) Regulations (Northern Ireland) 1999 (SR 1999 no 73)

Protection of Wrecks Act 1973 (1973 c. 33)

Roads (Scotland) Act 1984 (1984 c. 54)

Town and Country Planning Act 1990 (1990 c. 8)

The Town and Country Planning (Environmental Impact Assessment) (England and Wales) Regulations 1999 (SI 1999 no 293)

The Town and Country Planning (General Permitted Development) Order 1995 (SI 1995 no 418)

Town and Country Planning (Scotland) Act 1997 (1997 c. 8)

Water Resources Act 1991 (1991 c. 57)

Wildlife and Countryside Act 1981 (1981 c. 69)